OVER THE SEAWALL

U.S. Marines at Inchon

by Brigadier General Edwin H. Simmons, USMC (Ret)

Just three weeks away and there was still no approval from Washington for the Marines to land at Inchon on 15 September 1950. General of the Army Douglas MacArthur, determined to beat down the opposition to the landing, called a conference for late in the day, 23 August, at his headquarters in the Dai Ichi building in Tokyo.

Planning

As Commander in Chief, Far East (CinCFE), MacArthur considered himself empowered to conduct military operations more-or-less as he saw fit. But for an operation of the magnitude of Inchon and the resources it would require he needed approval from the highest level.

The Joint Chiefs of Staff (JCS), doubtful of the landing's chances of success, had sent out the Army Chief of Staff, General J. Lawton Collins, and the Chief of Naval Operations, Admiral Forrest P. Sherman, to review the situation directly with MacArthur. Now he would have to overcome their skeptical resistance. Collins was the JCS executive agent for the Far East Command and nominally higher in the chain-of-command than MacArthur—but only nominally. In World War I MacArthur was already a brigadier general when Collins was barely a captain. Now MacArthur had five stars and Collins four.

On this afternoon, First Lieutenant Alexander M. Haig's task was to lay out the pads of paper, pencils, and water glasses on the table of the sixth floor conference room. This done, he took his post seated in a straight-backed chair just outside the door. Haig, then the junior aide-de-camp to MacArthur's chief of staff, was destined to become, many years later, the Secretary of State.

The Marine Corps would have no voice at the meeting. The Corps had neither membership nor representation on the JCS. Admiral Sherman, not a strong champion of Marine Corps interests, was the service chief most directly concerned with the amphibious phase of the still tentative operation.

Opening Moves

Only two months before the meeting of MacArthur with Collins and Sherman, in the pre-dawn hours of 25 June, 25-year-old Lieutenant Haig, as duty officer at MacArthur's headquarters in Tokyo, received a phone call from the American ambassador in Seoul, John J. Muccio, that large formations of North Korean infantry had crossed the 38th Parallel. Haig informed his boss, Major General Edward M. "Ned" Almond, chief of staff of the Far East Command, who awakened MacArthur with the news. The United States was going to war.

Four days later, and a day after the fall of Seoul, MacArthur flew to Korea in the *Bataan*, to make a personal reconnaissance, taking with him Major General Almond. Korea stretched beneath them like a giant relief map. To the east of the Korean peninsula lay the Sea of Japan; to the west the Yellow Sea. The vulnerability of these two watery sides of the peninsula to a dominant naval power was not lost on a master strategist such as MacArthur. The *Bataan* landed at Suwon, 20 miles south of Seoul. MacArthur commandeered a jeep and headed north through, in his words, "the dreadful backwash of a defeated and dispersed army."

"Seoul was already in enemy hands," he wrote in his *Reminiscences* some years later. "The scene along the Han was enough to convince me that the defensive potential of South Korea had already been exhausted. The answer I had come to seek was there. I would throw occupation troops into this breach. I would rely upon strategic maneuver to overcome the great odds against me."

MacArthur returned to what he liked to call his "GHQ" in Tokyo, convinced that to regain the initiative the United States must use its amphibious capability and land behind the advancing North

---

ON THE COVER: *Using scaling ladders, Marines storm over the seawall at Inchon.* Department of Defense Photo (USMC) A3191

AT LEFT: *The mop-up at Inchon turned up a group of young North Koreans, left as harassing forces.* Photo courtesy of *Leatherneck* Magazine

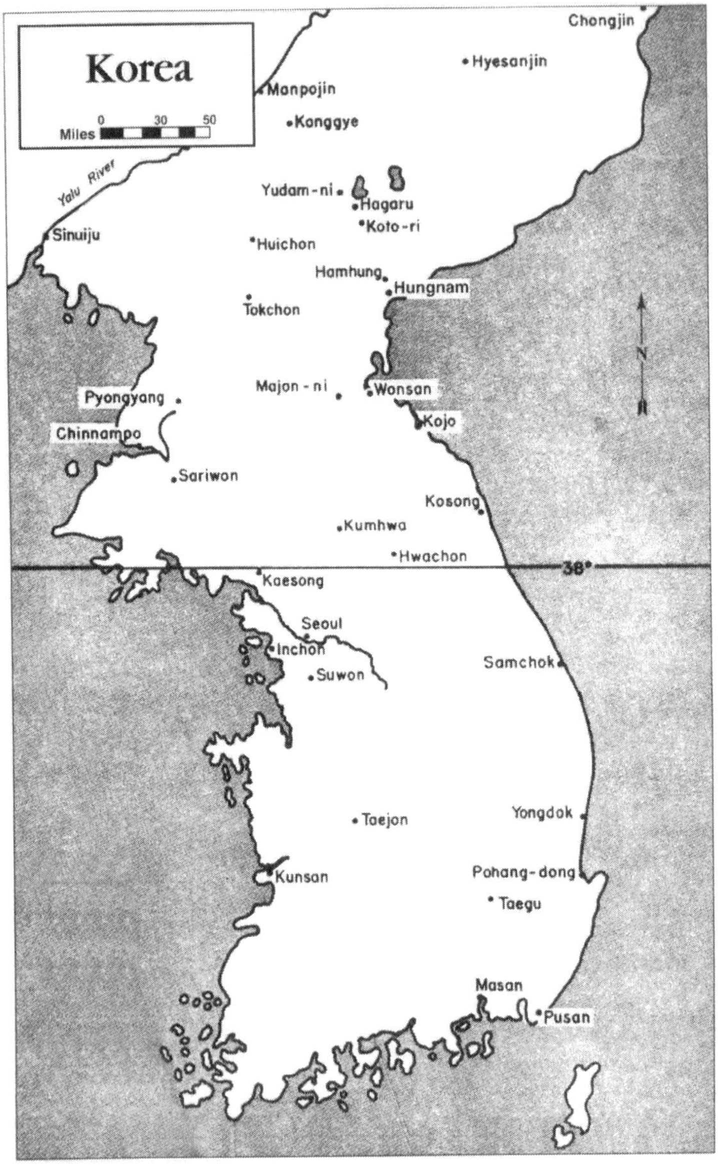

Koreans. He put his staff to work on a broad operational plan: two U.S. divisions would be thrown into the battle to slow the onrush of the North Korean People's Army (NKPA). A third division would land behind the NKPA and in a flanking attack liberate Seoul, the lost capital.

Unready Eighth Army

MacArthur had at his disposal in Japan the Eighth Army consisting of four divisions—the 7th, 24th, 25th, and the 1st Cavalry—all four at half-strength and under-trained. He began to move pieces of the 24th Division, rated at 65 percent combat-ready, to South Korea. His aim, he later said, was to trade space for time until a base could be developed at Pusan at the southern tip of the peninsula as a springboard for future operations.

Approval came from President Harry S. Truman for the imposition of a naval blockade and limited air operations. "The Air Force was under Lieutenant General George E. Stratemeyer, and the Navy under Vice Admiral C. Turner Joy, both able and efficient veterans of the war," wrote MacArthur.

But Vice Admiral Joy, as Commander Naval Forces, Far East, commanded virtually nothing. Vice Admiral Arthur D. "Rip" Struble, commander of the Seventh Fleet, a naval officer of considerable amphibious experience, reported not to Joy but to Admiral Arthur W. Radford who was both Commander in Chief, Pacific, and Commander in Chief, Pacific Fleet.

Lieutenant General George E. Stratemeyer commanded "FEAF" or Far East Air Forces. Subordinate to him were the Fifth Air Force in Japan, the Twentieth Air Force on Okinawa, and the Thirteenth Air Force in the Philippines.

Cates Offers Marines

Back in Washington, D.C., during the first hectic days after the North Korean invasion, the Commandant of the Marine Corps, General Clifton B. Cates, was not invited to attend the high-level meetings being held in the Pentagon. After four days of waiting, Cates drove to the Pentagon and, in his words, "kind of forced my way in."

"We were fighting for our existence," said General Lemuel C. Shepherd, Jr., who followed Cates as Commandant. "Sherman and the rest of these fellows wanted to keep us seagoing Marines, with a battalion landing team being the biggest unit we were supposed to have . . . . Everybody was against the Marine Corps at that time. Secretary of Defense Louis A. Johnson, always nagging, Truman hostile, and Cates carried that load all by himself and did it well."

Cates saw Admiral Sherman and told him the Marines could immediately deploy to Korea a brigade consisting of a regimental combat team and an aircraft group.

"How soon can you have them ready?" Sherman asked dubiously.

"As quickly as the Navy gets the ships," shot back Cates.

Sherman, overwhelmed perhaps by higher priorities, dallied two days before sending a back-channel message to Admiral Joy, asking him to suggest to MacArthur that he request a Marine air-ground brigade. MacArthur promptly made the request and on 3 July the JCS approved the deployment.

Cates did not wait for JCS approval. Formation of the 1st Provisonal Marine Brigade had already begun with troops stripped out of the half-strength 1st Marine Division. In four days' time—on 6 July—the brigade began to load out at San Diego for the Far East.

Several months before the breakout of war, MacArthur had requested amphibious training for his occupation troops. Troop Training Unit, Amphibious Training Command, Pacific Fleet, had been formed in 1943 for just such a purpose. Colonel Edward H. Forney, with Mobile Training Team Able and accompanied by an Air and Naval Gunfire Liaison Company (ANGLICO) training team, arrived in April 1950. A regiment in each of MacArthur's four divisions was to be amphibiously trained. Navy partner in the training would be Amphibious Group One (PhibGruOne) under Rear Admiral James H. Doyle.

A few days before the outbreak of the war Brigadier General William S. Fellers, commanding general of the Troop Training Unit, came out to Japan to inspect the progress being made by Forney and his team. Fellers and Forney were at a Fourth of July party being given by the American colony in Tokyo when an urgent message required their immediate presence at "GHQ." They arrived at the Dai Ichi—a tall building that had escaped the World War II bombing because the Imperial Palace was immediately across the way—to find a planning conference in progress with Almond at the helm. They learned that MacArthur had advanced the concept of a landing at Inchon, to be called Operation Bluehearts and to be executed on 22 July by the 1st Cavalry Division—and the 1st Provisional Marine Brigade, if the latter could be gotten there in time. Next day Colonel Forney became the G-5 (Plans) of the 1st Cavalry, one of MacArthur's favorite divisions.

Shepherd Meets with MacArthur

Three days after the interrupted Fourth of July party, Lemuel Shepherd, just promoted to lieutenant general and installed as Commanding General, Fleet Marine Force, Pacific, left Hawaii for Tokyo, accompanied by his operations officer, Colonel Victor H. Krulak. Shepherd had been urged to go to Tokyo by Admiral Radford, a good friend of the Marines, "to see MacArthur and find out what all this thing is about."

Shepherd saw his mission as being first to ensure that the 1st Provisional Marine Brigade was used as an integrated air-ground team and, second, to explore prospects for the use of additional Marine Corps forces.

"Having been with the 4th Brigade in France, I had learned that a Marine unit in an Army division is not good for the Corps," said Shepherd years later. Enroute to Tokyo he made up his mind that he was going to push for a Marine division to be sent to Korea.

General Shepherd met with Admiral Joy and General Almond on 9 July, and next day, accompanied by Colonel Krulak, saw MacArthur himself. He told them that the only hope for an early reversal of the disastrous situation was an amphibious assault against the enemy's rear.

"Here I was," said Shepherd later, "recommending that a Marine division be sent to Korea, and the Commandant didn't know anything about what I was doing."

MacArthur recalled to Shepherd the competence of the 1st Marine Division when it had been under his command during the Cape Gloucester operation at Christmas time in 1943. Shepherd had then been the assistant division commander. MacArthur went to his wall map, stabbed at the port of Inchon with the stem of his corncob pipe, and said: "If I only had the 1st Marine Division under my command again, I would land them here and cut off the North Korean armies from their logistic support and cause their withdrawal and annihilation."

Shepherd answered that if MacArthur could get JCS approval for the assignment of the 1st Marine Division, he could have it ready by mid-September. MacArthur told Shepherd to draft for his release a message to the JCS asking for the division.

Bluehearts, which would have used the 1st Cavalry Division, was abruptly cancelled. Planning in Tokyo, under Brigadier General Edwin K. "Pinky" Wright, USA, and his Joint Strategic Plans and Operations Group (JSPOG), shifted to an amphibious operation in September.

Under the U.N. Flag

On that same busy 10 July, MacArthur's mantle of authority was embroidered with a new title—Commander in Chief, United Nations Command or "CinCUNC." From then on operations in Korea and surrounding waters would be fought under the light-blue-and-white flag of the United Nations.

The sailing of the 1st Provisional Marine Brigade from San Diego began on 12 July. Core of the ground element was the 5th Marines; the air element was Marine Aircraft Group 33. Filling the brigade had gutted both the 1st Marine Division and the 1st Marine Aircraft Wing.

General Cates was in San Diego to see the Marines off. His long cigarette holder was famous; not many Marines knew that he used it because gas in World War I had weakened his lungs. General Shepherd was also on the dock and it gave him the opportunity to discuss with Cates his promise to MacArthur of a full division. Could the 1st Marine Division be assembled and made ready in such a short time?

"I don't know," said Cates dubiously; it would drain the Marine Corps completely.

"Clifton," said Shepherd simply, "you can't let me down."

Visitors from Washington

In Tokyo, where it was already 13 July, MacArthur was meeting with visitors from Washington—Army General Collins and General Hoyt S. Vandenberg, chief of staff of the newly independent Air Force. Also present were Admiral Radford, General Almond, and Lieutenant General Walton H. Walker. It had just been announced that Walker was shifting his flag from Japan to Korea, and the Eighth Army would become the Eighth U.S. Army in Korea, which yielded the acronym "EUSAK." MacArthur explained his reasons for cancelling Bluehearts and said that he had not yet chosen a new target date or location for an amphibious strike, but favored Inchon.

As soon as the meeting was over, Collins and Walker flew to Korea, where Walker opened a field headquarters at Taegu for his Eighth Army. Collins spent only an hour on the ground and did not leave the airport before returning to Tokyo.

Next day, the 14th, he was briefed by General Almond and Admiral Doyle, who had commanded Amphibious Group One since January. Before that for two years Doyle had headed the Amphibious Training Command, Pacific Fleet. During World War II he served on the staff of Amphibious Force, South Pacific.

Collins questioned the feasibility of landing at Inchon. Doyle said that it would be difficult but could be done. Before leaving Tokyo, Collins assured MacArthur that he would endorse the sending of a full-strength Marine division.

Earlier, during the planning for Operation Bluehearts, Doyle had expressed reservations over the use of the 1st Cavalry Division because it was not amphibiously trained. His relations with Almond were strained. He thought Almond arrogant and dictatorial and a person who "often confused himself with his boss."

Lieutenant Haig, Almond's aide and the keeper of his war diary, found his chief "volcanic" in personality, "brilliant" but "irascible," and, with all that, a "phenomenally gifted soldier." Almond, like his idol, General George S. Patton, Jr., designed his own uniforms and wore a pistol on a leather belt adorned with a huge crested buckle. He did this, he said, so as to be easily recognized by his troops.

General Walker, a tenacious man who deserved his nickname "Bulldog" (although he was "Johnnie" Walker to his friends), continued the piecemeal buildup of the Eighth Army. All of the 24th Division was committed by 7 July. The 25th Division completed its move from Japan on 14 July.

Tactical Air Control Problems

The 1st Cavalry Division was in process of loading out from Japan in Doyle's PhibGruOne when Bluehearts was cancelled in favor of an unopposed landing on 18 July at Pohang-dong, a port some 60 air miles northeast of Pusan. Plans developed for Bluehearts by both PhibGruOne and 1st Cavalry Division were used for the operation. For this non-hostile landing the Navy insisted on control of an air space 100 miles in diameter circling the landing site. This Navy requirement for control of air traffic over the objective area conflicted with Air Force doctrine which called for Air Force control of all tactical aircraft in the theater of operations.

Lieutenant General Earle E. "Pat" Partridge, whose Fifth Air Force Joint Operations Center was in Taegu side-by-side with Walker's Eighth Army headquarters, protested the Navy requirement that would have caused him to vacate the control of air over virtually all of the Pusan Perimeter. This began a doctrinal dispute involving the tactical control of air that would continue for the rest of the war.

Major General Oliver P. Smith

Oliver Prince Smith did not fill the Marine Corps "warrior" image. He was deeply religious, did not drink, seldom raised his voice in anger, and almost never swore. Tall, slender, and white-haired, he looked like a college professor is supposed to look and seldom does. Some of his contemporaries thought him pedantic and a bit slow. He smoked a pipe in a meditative way, but when his mind was made up he could be as resolute as a rock. He always commanded respect and, with the passage of years, that respect became love and devotion on the part of those Marines who served under him in Korea. They came to know that he would never waste their lives needlessly.

As commanding general of the 1st Marine Division, Smith's feud with the mercurial commander of X Corps, Major General Edward M. Almond, USA, would become the stuff of legends.

No one is ever known to have called him "Ollie." To his family he was "Oliver." To his contemporaries and eventually to the press, which at first tended to confuse him with the controversial Holland M. "Howlin' Mad" Smith of World War II, he was always "O. P." Smith. Some called him "the Professor" because of his studious ways and deep reading in military history.

Born in Menard, Texas, in 1893, he had by the time of America's entry into the First World War worked his way through the University of California at Berkeley, Class of 1916. While a student at Berkeley he qualified for a commission in the Army Reserve which he exchanged, a week after America's entry into the war on 6 April 1917, for the gold bars of a Marine Corps second lieutenant.

Department of Defense Photo (USMC) A8377

The war in Europe, where the Marines gained international fame, passed him by; he spent the war years in lonely exile with the garrison on Guam. Afterward, in the 1920s, he followed an unremarkable sequence of duty, much like that of most lieutenants and captains of the time: barracks duty at Mare Island, sea duty in the *Texas*, staff duty at Headquarters Marine Corps, and a tour with the Gendarmerie d'Haiti.

From June 1931 to June 1932, he attended the Field Officer's Course at Fort Benning. Next came a year at Quantico, most of it spent as an instructor at the Company Officer's Course. He was assigned in 1934 to a two-year course at the Ecole Superieur la de Guerre in Paris, then considered the world's premier school for rising young officers. Afterwards he returned to Quantico for more duty as an instructor.

The outbreak of World War II in 1939 found him at San Diego. As commanding officer of the 1st Battalion, 6th Marines, he went to Iceland in the summer of 1941. He left the regiment after its return to the States, for duty once again at Headquarters in Washington. He went to the Pacific in January 1944 in time to command the 5th Marines during the Talasea phase of the Cape Gloucester operation. He was the assistant commander of the 1st Marine Division during Peleliu and for Okinawa was the Marine Deputy Chief of Staff of the Tenth Army.

After the war he was the commandant of Marine Corps Schools and base commander at Quantico until the spring of 1948 when he became the assistant commandant and chief of staff at Headquarters. In late July 1950, he received command of the 1st Marine Division, destined for Korea, and held that command until May 1951.

After Inchon and Seoul, a larger, more desperate fight at Chosin Reservoir was ahead of him. In early 1951, the 1st Marine Division was switched from Almond's X Corps to Major General Bryant E. Moore's IX Corps. Moore died of a heart attack on 24 February 1951 and, by seniority, O. P. Smith became the corps commander. Despite his experience and qualifications, he held that command only so long as it took the Army to rush a more senior general to Korea.

O. P. Smith's myriad of medals included the Army Distinguished Service Cross and both the Army and the Navy Distinguished Service Cross for his Korean War Service.

On his return to the United States, he became the commanding general of the base at Camp Pendleton. Then in July 1953, with a promotion to lieutenant general, moved to the East Coast to the command of Fleet Marine Force, Atlantic, with headquarters at Norfolk, Virginia. He retired on 1 September 1955 and for his many combat awards was promoted to four-star general. He died on Christmas Day 1977 at his home in Los Altos Hills, California, at age 81.

Joint Chiefs Reluctant

Returning to Washington, Collins briefed his fellow chiefs on 15 July. He gave them the broad outlines of MacArthur's planned amphibious assault, but expressed his own doubts based on his experience in the South Pacific and at Normandy.

The JCS chairman, General of the Army Omar N. Bradley, thought it "the riskiest military proposal I ever heard of." In his opinion, MacArthur should be concentrating on the dismal immediate situation in South Korea rather than dreaming up "a blue sky scheme like Inchon." Bradley wrote later: "because Truman was relying on us to an extraordinary degree for military counsel, we determined to keep a close eye on the Inchon plan and, if we felt so compelled, finally cancel it."

The JCS agreed that the 1st Marine Division should be brought up to strength, but stopped short of committing it to the Far East. On 20 July, the Joint Chiefs informed MacArthur that the 1st Marine Division could not be combat ready until December. MacArthur erupted: the 1st Marine Division was "absolutely vital" to the plan being developed, under the codename Chromite, by General Wright's group. A draft, circulated at CinCFE headquarters on 23 July, offered three alternatives:

> Plan 100-B: A landing at Inchon on the west coast.
>
> Plan 100-C: A landing at Kunsan on the west coast.
>
> Plan 100-D: A landing at Chunmunjin-up on the east coast.

MacArthur's mind was now fully set on Inchon. He informed Collins, in his capacity as executive agent for the JCS, that lacking the Marine division, he had scheduled an amphibious assault at Inchon in mid-September to be executed by the 5th Marines and the 2d Infantry Division in conjunction with an attack northward by the Eighth Army. His message caused the chiefs to initiate a hurried teletype conference with MacArthur on 24 July. MacArthur prevailed and on the following day, 25 July, the chiefs finally approved MacArthur's repeated requests for the 1st Marine Division.

Gen Oliver P. Smith Collection, Marine Corps Research Center

MajGen Oliver P. Smith, left, assumed command of the 1st Marine Division at Camp Pendleton on 26 July 1950. Col Homer L. Litzenberg, Jr., right, arrived on 16 August with orders to reactivate the 7th Marines and have it ready for sailing by 3 September.

A New CG

Late in the afternoon of 25 July, Major General Oliver P. Smith arrived from Washington and checked in at the Carlsbad Hotel in Carlsbad, California. He was to take command of the 1st Marine Division at nearby Camp Pendleton on the following day. He phoned Brigadier General Harry B. Liversedge, the base commander and acting division commander, to let him know that he had arrived. Liversedge said that

Colonel Lewis B. "Chesty" Puller

Gen Oliver P. Smith Collection

The younger Marines in the 1st Marines were ecstatic when they learned their regiment was going to be commanded by the legendary "Chesty" Puller. Older officers and non-commissioned officers in the regiment were less enthusiastic. They remembered the long casualty list the 1st Marines had suffered at Peleliu while under Colonel Puller's command. His style was to lead from the front, and, when he went into Korea, he already had an unprecedented four Navy Crosses.

Born in 1898, Puller had grown up in Tidewater Virginia where the scars of the Civil War were still unhealed and where many Confederate veterans were still alive to tell a young boy how it was to go to war. Lewis (which is what his family always called him) went briefly to Virginia Military Institute but dropped out in August 1918 to enlist in the Marines. To his disappointment, the war ended before he could get to France. In June 1919, he was promoted to second lieutenant and then, 10 days later, with demobilization was placed on inactive duty. Before the month was out he had reenlisted in the Marines specifically to serve as a second lieutenant in the Gendarmerie d'Haiti. Most of the officers in the Gendarmerie were white Marines; the rank and file were black Haitians. Puller spent five years in Haiti fighting "Caco" rebels and making a reputation as a bush fighter.

He returned to the States in March 1924 and received his regular commission in the Marine Corps. During the next two years he did barracks duty in Norfolk, attended Basic School in Philadelphia, served in the 10th Marines at Quantico, and had an unsuccessful try at aviation at Pensacola. Barracks duty for two years at Pearl Harbor followed Pensacola. Then in 1928 he was assigned to the Guardia Nacional of Nicaragua. Here in 1930 he won his first Navy Cross. First Lieutenant Puller, his citation reads, "led his forces into five successive engagements against superior numbers of armed bandit forces."

He came home in July 1931 to the year-long Company Officers Course at Fort Benning. That taken, he returned to Nicaragua for more bandit fighting and a second Navy Cross, this time for taking his patrol of 40 Nicaraguans through a series of ambushes, in partnership with the almost equally legendary Gunnery Sergeant William A. "Iron Man" Lee.

Now a captain, Puller came back to the West Coast in January 1933, stayed a month, and then left to join the Legation Guard at Peiping. This included command of the fabled "Horse Marines." In September 1934, he left Peiping to become the commanding officer of the Marine detachment on board the *Augusta*, flagship of the Asiatic Fleet.

In June 1936, he came to Philadelphia to instruct at the Basic School. His performance as a tactics instructor and on the parade ground left its mark on the lieutenants who would be the captains, majors, and lieutenant colonels in the world war that was coming.

In June 1939, he went back to China, returning to the *Augusta* to command its Marines once again. A year later he left the ship to join the 4th Marines in Shanghai. He returned to the United States in August 1941, four months before the war began, and was given command of the 1st Battalion, 7th Marines, at Camp Lejeune. He commanded (he would say "led") this battalion at Guadalcanal and won his third Navy Cross for his successful defense of a mile-long line on the night of 24 October 1942. The fourth Navy Cross came for overall performance, from 26 December 1943 to 19 January 1944, at Cape Gloucester as executive officer of the 7th Marines. In February 1944, he took command of the 1st Marines and led it in the terrible fight at Peleliu in September and October.

Afterwards, he came back to command the Infantry Training Regiment at Camp Lejeune. Next he was Director of the 8th Marine Corps Reserve District with headquarters in New Orleans, and then took command of the Marine Barracks at Pearl Harbor. From here he hammered Headquarters to be given command, once again, of his old regiment, the 1st Marines.

After Inchon, there was to be a fifth Navy Cross, earned at the Chosin Reservoir. In January 1951, he received a brigadier general's stars and assignment as the assistant division commander. In May, he came back to Camp Pendleton to command the newly activated 3d Marine Brigade which became the 3d Marine Division. He moved to the Troop Training Unit, Pacific, on Coronado in June 1952 and from there moved east, now with the two stars of a major general, to Camp Lejeune to take command of the 2d Marine Division in July 1954. His health began to fail and he was retired for disability on 1 November 1955. From then until his death on 11 October 1971 at age 73 he lived in the little town of Saluda in Tidewater Virginia.

he had just received a tip from Washington that the division was to be brought to war strength and sail to the Far East by mid-August. Both Liversedge and Smith knew that what was left of the division was nothing more than a shell.

Smith took command the next day, 26 July. He had served in the division during World War II, commanding the 5th Marines in its Talasea landing at New Britain and was the assistant division commander at Peleliu. Only 3,459 Marines remained in the division at Camp Pendleton, fewer men than in a single full-strength regiment.

When the Joint Chiefs asked General Cates how he planned to bring the 1st Marine Division up to war strength, he had ready a two-pronged plan. Plan A would provide three rifle companies and replacements to the brigade already deployed. Plan B would use Reserves to fill up the division. Essential to the filling out of the 1st Marine Division—and the 1st Marine Aircraft Wing as well—was the mobilization of the Marine Corps Reserve. "Behind every Marine regular, figuratively speaking," wrote official historians Lynn Montross and Captain Nicholas A. Canzona, "stood two reservists who were ready to step forward and fill the gaps in the ranks."

The 33,527 Marines in the Organized Reserve in 1950 were scattered across the country in units that included 21 infantry battalions and 30 fighter squadrons. Virtually all the officers and noncommissioned officers had World War II experience, but the ranks had been filled out with youngsters, many of whom did not get to boot camp. Subsequent reserve training had included both weekly armory "drills" and summer active duty. Someone wryly decided they could be classified as "almost combat ready."

Behind the Organized Reserve was the Volunteer Marine Corps Reserve—90,044 men and women, most of them veterans, but with no further training after their return to civilian life. President Truman, with the sanction of Congress, authorized the call-up of the Marine Corps Reserve on 19 July. An inspired public information officer coined the phrase, "Minute Men of 1950."

On 26 July, the day following JCS approval of the 1st Marine Division's deployment, a courier arrived at Camp Pendleton from Washington with instructions for Smith in his fleshing-out of the 1st Marine Division: ground elements of the 1st Provisional Marine Brigade would re-combine with the division upon its arrival in the Far East; units of the half-strength 2d Marine Division at Camp Lejeune, North Carolina, would be ordered to Camp Pendleton and re-designated as 1st Marine Division units; all possible regulars would be stripped out of posts and stations and ordered to the division; and gaps in the ranks would be filled with individual Reserves considered to be at least minimally combat-ready.

Eighth Army Withdraws to Pusan

In Korea, at the end of July, Walker ordered the Eighth Army to fall back behind the Naktong River, the new defensive line forming the so-called "Pusan Perimeter." Both flanks of the Eighth Army were threatened. In light of this deteriorating situation, the Joint Chiefs asked MacArthur if he still planned an amphibious operation in September. An unperturbed MacArthur replied that "if the full Marine Division is provided, the chances to launch the movement in September would be excellent."

Reinforcements for Walker's Eighth Army began arriving directly from the United States, including the 1st Provisional Marine Brigade which debarked at Pusan on 3 August.

In Tokyo, General Stratemeyer became agitated when he learned that the 1st Provisional Marine Brigade, as an integrated air-ground team, intended to retain mission control of its aircraft. An uneasy compromise was reached by which the Marines were to operate their two squadrons of carrier-based Vought F4U Corsairs with their own controllers under the general coordination of Partridge's Fifth Air Force.

Reserve Comes to Active Duty

The first reservists to reach Pendleton—the 13th Infantry Company from Los Angeles, the 12th Amphibian Tractor Company of San Francisco, and the 3d Engineer Company from Phoenix—arrived on 31 July. Elements of the 2d Marine Division from Camp Lejeune began their train journey the same day. In that first week, 13,703 Marines joined the division.

On 4 August, the Commandant ordered the reactivation of the 1st Marines and 7th Marines. Both regiments had been part of the 1st Marine Division in all its World War II campaigns. The 1st Marines was activated that same day under command of the redoubtable Colonel Lewis B. "Chesty" Puller, who, stationed at Pearl Harbor as commanding officer of the Marine Barracks, had pestered Headquarters Marine Corps and General Smith with demands that he be returned to the command of the regiment he had led at Peleliu. By 7 August, the strength of the 1st Marine Division stood at 17,162.

The experiences of Lieutenant Colonel Thomas L. Ridge's 1st

Battalion, 6th Marines, were typical of the buildup being done at a dead run. Ridge had just taken command of the battalion. A crack rifle and pistol shot, he had spent most of World War II in intelligence assignments in Latin America, but in late 1944 was transferred to Fleet Marine Force, Pacific, in time for staff duty for Iwo Jima and Okinawa. As an observer at Okinawa he was twice wounded.

Ridge's battalion, barely returned to Camp Lejeune from six months deployment to the Mediterranean, traveled by ancient troop train to Camp Pendleton where it became the 3d Battalion of the reactivated 1st Marines. In about 10 days, the two-element, half-strength battalion expanded into a three-element, full-strength battalion. The two rifle companies

Major General Field Harris

During the course of the Korean War, Major General Field Harris would suffer a grievous personal loss. While he served as Commanding General, 1st Marine Aircraft Wing, his son, Lieutenant Colonel William F. Harris, was with the 1st Marine Division, as commanding officer of 3d Battalion, 7th Marines, at the Chosin Reservoir. The younger Harris' battalion was the rear guard for the breakout from Yudam-ni. Later, between Hagaru-ri and Koto-ri, Harris disappeared and was posted as missing in action. Later it was determined that he had been killed.

Field Harris—and he was almost always called that, "Field-Harris," as though it were one word—belonged to the open cockpit and silk scarf era of Marine Corps aviation. Born in 1895 in Versailles, Kentucky, he received his wings at Pensacola in 1929. But before that he had 12 years seasoning in the Marine Corps.

He graduated from the Naval Academy in March 1917 just before America's entry into World War I. He spent that war at sea in the *Nevada* and ashore with the 3d Provisional Brigade at Guantanamo, Cuba.

In 1919 he went to Cavite in the Philippines. After three years there, he returned for three years in the office of the Judge Advocate General in Washington. While so assigned he graduated from the George Washington University School of Law. Then came another tour of sea duty, this time in the *Wyoming*, then a year as a student at Quantico, and flight training at Pensacola. His new gold wings took him to San Diego where he served in a squadron of the West Coast Expeditionary Force.

He attended the Air Corps Tactical School at Langley Field, Virginia, after which came shore duty in Haiti and sea duty in the carrier *Lexington*. In 1935, he joined the Aviation Section at Headquarters, followed by a year in the Senior Course at the Naval War College in Newport, Rhode Island. In August 1941, he was sent to Egypt from where, as assistant naval attache, he could study the Royal Air Force's support of Britain's Eighth Army in its desert operations.

After Egypt and United States entry into the war, he was sent to the South Pacific. In the Solomons, he served successively as Chief of Staff, Aircraft, Guadalanal; Commander, Aircraft, Northern Solomons; and commander of air for the Green Island operation. Each of these

Department of Defense Photo (USMC) A310952

three steps up the chain of islands earned him a Legion of Merit. After World War II, he became Director of Marine Aviation in the Office of the Chief of Naval Operations (and received a fourth Legion of Merit). In 1948 he was given command of Aircraft, Fleet Marine Force, Atlantic. A year later he moved to El Toro, California, for command of Aircraft, Fleet Marine, Pacific, with concomitant command of the 1st Marine Aircraft Wing.

His Korean War service was rewarded with both the Army's and the Navy's Distinguished Service Medal. On his return to the United States in the summer of 1951, he again became the commanding general of Air, Fleet Marine Force, Atlantic. He retired in July 1953 with an advancement to lieutenant general because of his combat decorations, a practice which is no longer followed. He died in 1967 at age 72.

in the battalion each numbering about 100 men were doubled in size with a third rifle platoon added. A third rifle company was activated. The weapons company had no heavy machine gun platoon and only two sections in its antitank assault and 81mm mortar platoons. A heavy machine gun platoon was created and third sections were added to the antitank assault and 81mm mortar platoons. World War II vintage supplies and equipment came in from the mobilization stocks stashed away at the supply depot at Barstow, California—sufficient in quantity, poor in quality. The pressure of the unknown D-Day gave almost no time for unit shake-down and training.

Simultaneously with the ground unit buildup, Reserve fighter and ground control squadrons were arriving at El Toro, California, to fill out the skeleton 1st Marine Aircraft Wing. The wing commander, Major General Field Harris, Naval Academy 1917, and a naval aviator since 1929, had served in the South Pacific in World War II. More recently he had been Director of Aviation at Marine Corps Headquarters. He was one of those prescient senior Marines who foresaw a future for helicopters in amphibious operations.

Department of Defense Photo (USMC) A20115

MajGen Field Harris and a portion of the 1st Marine Aircraft Wing staff arrive at Barber's Point in Hawaii in early September enroute to the Far East. Leaving the Marine transport are, from left, Col Edward C. Dyer, Col Boeker C. Batterton, Col William G. Manley, and Gen Harris.

7th Infantry Division and KATUSA

In parallel actions, MacArthur on 4 August ordered Walker to rebuild the Army's 7th Infantry Division—the last division remaining in Japan—to full strength by 15 September. The division had been reduced to less than half-strength by being repeatedly culled for fillers for the three divisions already deployed to Korea. Until MacArthur's directive, the division was not scheduled to be up to strength until 1 October and not ready for amphibious operations until 1951. Now, the division was to get 30 percent of all replacements arriving from the United States. Moreover, a week later, on 11 August, MacArthur directed Walker to send 8,000 South Korean recruits to fill out the division.

The first of 8,600 Korean replacements, straight out of the rice paddies of South Korea and off the streets of Pusan, began arriving by ship at Yokohama a few days later. This infusion of raw untrained manpower, called "KATUSA"—Korean Augmentation of the U.S. Army—arrived for the most part in baggy white pants, white jackets, and rubber shoes. In three weeks they had to be clothed, equipped, and made into soldiers, including the learning of rudimentary field sanitation as well as rifle practice. The "buddy system" was employed—each Korean recruit was paired off with an American counterpart.

Major General David G. Barr, the 7th Infantry Division's commander, had been chief of staff of several commands in Europe during World War II. After the war he had headed the Army Advisory Mission in Nanking, China. He now seemed a bit old and slow, but he knew Chinese and the Chinese army.

1st Marine Division Loads Out

Loading out of the 1st Marine Division from San Diego began on 8 August. That same day, General Fellers, back from Japan, told Smith that the division would be employed in Korea between 15

Colonel Homer L. Litzenberg, Jr.

His troops called him "Litz the Blitz" for no particular reason except the alliteration of sound. He had come up from the ranks and was extraordinarily proud of it. Immediately before the Korean War began he was in command of the 6th Marines at Camp Lejeune, very much interested in his regimental baseball teams, and about to turn over the command to another colonel. When war came he was restored to command of the regiment and sadly watched his skeleton battalions depart for Camp Pendleton to form the cadre for the re-activated 1st Marines. This was scarcely done when he received orders to re-activate the 7th Marines on the West Coast.

Litzenberg was a "Pennsylvania Dutchman," born in Steelton, Pennsylvania, in 1903. His family moved to Philadelphia and, after graduating from high school and two years in the National Guard, he enlisted in the Marine Corps in 1922. Subsequent to recruit training at Parris Island, he was sent to Haiti. In 1925 he became a second lieutenant. East Coast duty was followed by expeditionary service in Nicaragua in 1928 and 1929, and then by sea service in a string of battleships—*Idaho*, *Arkansas*, *Arizona*, *New Mexico*—and the cruiser *Augusta*. After graduating from the Infantry School at Fort Benning in 1933, he had two years with a Marine Reserve battalion in Philadelphia. Next came two years on Guam as aide to the governor and inspector-instructor of the local militia. He came home in 1938 to serve at several levels as a war planner.

When World War II came, he was sent, as a major, to England to serve with a combined planning staff. This took him to North Africa for the amphibious assault of Casablanca in November 1942. He came home to form and command the 3d Battalion, 24th Marines, in the new 4th Marine Division, moving up to regimental executive officer for the assault of Roi-Namur in the Marshalls. He then went to the planning staff of the V Amphibious Corps for Saipan and Tinian.

After the war he went to China for duty with the Seventh Fleet and stayed on with Naval Forces Western Pacific. He came home in 1948 and was given command of the 6th Marines in 1949.

Department of Defense Photo (USMC) A4718

After Inchon, he continued in command of the 7th Marines through the battles of Seoul, Chosin Reservoir, and the Spring Offensive, coming home in April 1951. Soon promoted to brigadier general and subsequently to major general, he had many responsible assignments including assistant command of the 3d Marine Division in Japan, Inspector General of the Marine Corps, command of Camp Pendleton, and command of Parris Island. He returned to Korea in 1957 to serve as senior member of the United Nations component negotiating at Panmunjom. At the end of the year he came back for what would be his last assignment, another tour of duty as Inspector General.

He retired in 1959, with an elevation to lieutenant general because of his combat awards that included a Navy Cross, a Distinguished Service Cross, and three Silver Stars. He died in the Bethesda Naval Hospital on 27 June 1963 at age 68 and was buried in Arlington National Cemetery with full military honors.

and 25 September.

Much of the heavy equipment to be loaded arrived at dockside from the Barstow supply depot with no time for inspection. General Shepherd arrived on 13 August to observe and encourage, joined next day by General Cates. Puller's 1st Marines sailed from San Diego on 14 August, 10 days after activation. The Navy had very little amphibious shipping on the West Coast, and much of the division and its gear had to be lifted by commercial shipping.

Among the pressing matters discussed by Smith with his superiors Cates and Shepherd was the reactivation of the 7th Marines. Nucleus of the 7th Marines would be the skeleton 6th Marines, which had already lost two battalions to the 1st Marines. The 3d Battalion, 6th Marines, a half-strength peacetime battalion with pieces scattered around the Mediterranean, became the 3d Battalion, 7th Marines, with orders to proceed to Japan by way of the Suez Canal. Fillers for the battalion and a completely new third rifle company would have to come from Camp Pendleton.

What was left of the 6th Marines arrived at Pendleton on 16 August. The 7th Marines activated the next day. Colonel Homer L. "Litz the Blitz" Litzenberg, Jr., a mercurial man who had commanded the 6th Marines at Camp Lejeune, continued as commanding officer of the 7th Marines with orders to embark his regiment not later than 3 September.

Joint Chiefs Have a Problem

Although the National Defense Act of 1947 was in effect, the relationship of the Joint Chiefs to the theater commanders was not too clear. As a theater commander MacArthur had broad leeway in his actions. The JCS faced the Hobson's choice of asking MacArthur no questions and making no challenges, or exerting their capacity as the principal advisors to President Truman in his role.

The Joint Chiefs held an intensive series of briefings in the White House on 10 August, culminating in an afternoon meeting with the National Security Council. President Truman was told that a war-strength Marine division was being assembled for service in Korea. Admiral Sherman assured the President, however, that the JCS would have to pass on MacArthur's plans for an amphibious operation.

On 12 August, MacArthur issued CinCFE Operations Plan 100-B, specifically naming Inchon-Seoul as the objective area. No copy of this plan was sent to the JCS.

O. P. Smith Departs Pendleton

General Smith sent off the first echelon of his division headquarters by air on 16 August. Two days later he closed his command post at Camp Pendleton and left by air for Japan. Delayed by shipping shortages, outloading of a third of Smith's division—essentially the reinforced 1st Marines—was completed on 22 August. In all, 19 ships were employed.

Following close behind, Litzenberg beat by two days the embarkation date given him by Smith. The 7th Marines, filled up with regulars pulled away from posts and stations and reservists, sailed from San Diego on 1 September.

Marine Versus Air Force Close Support

General Stratemeyer, MacArthur's Air Force component commander, apparently first heard of the possibility of an Inchon landing on 20 July. His first action was to instruct his staff to prepare a small command group with which he could accompany MacArthur on the operation. Almost a month later, on 14 August, MacArthur discussed the proposed landing with Stratemeyer, pointing out that Kimpo Airfield, just west of the Han River from Seoul, was the best in Korea. MacArthur emphasized that the airfield must be quickly rehabilitated from any battle damage and put to use.

By then news stories were appearing that compared Fifth Air Force support of the Eighth Army unfavorably with the close air support being provided the Marine

USS Mount McKinley *(AGC7) was the command center afloat for the Inchon landing. It also served as a floating hotel for the large number of VIPs who were in Gen Douglas MacArthur's official party or were simply passing through.*
National Archives Photo (USN) 80-G-424523

brigade by its organic squadrons. On 23 August, Stratemeyer sent a memorandum to MacArthur stating that the news stories were another step "in a planned program to discredit the Air Force and the Army and at the same time to unwarrantedly enhance the prestige of the Marines." He pointed out that the Marine squadrons, operating from two aircraft carriers, were supporting a brigade of about 3,000 Marines on a front that could be measured in yards as compared to the Fifth Air Force which had to supply close air support for a front of 160 miles.

General Walker, collocated at Taegu with General Partridge, pulled the rug out from under General Stratemeyer's doctrinal concerns and contentions of unfairness, by commenting officially: "Without the slightest intent of disparaging the support of the Air Force, I must say that I, in common with the vast majority of officers of the Army, feel strongly that the Marine system of close air support has much to commend it . . . . I feel strongly that the Army would be well advised to emulate the Marine Corps and have its own tactical aviation."

National Archives Photo (USN) 80-G-422492

Gen Douglas MacArthur, center, greets Gen J. Lawton Collins, Chief of Staff, U.S. Army, and Adm Forrest P. Sherman, Chief of Naval Operations, upon their arrival in Tokyo on 21 August 1950. A critical conference would be held two days later at which MacArthur would have to convince these two members of the Joint Chiefs of Staff that a landing at Inchon was feasible.

Top Brass Gathers in Tokyo

General Collins and Admiral Sherman—the latter had not been to Korea before—made a quick visit on 22 August to Walker's Eighth Army headquarters at Taegu. Collins found Walker "too involved in plugging holes in his leaky front to give much thought to a later breakout." On the morning of 23 August, Collins accompanied Walker on a visit to all U. S. division commanders and the Marine brigade commander, Brigadier General Edward A. Craig. Collins found these field commanders confident but weary. Collins then returned to Tokyo for the crucial conference at which MacArthur must overcome JCS reservations concerning the Inchon landing.

Major General Smith arrived at Haneda airport in Japan on 22 August and was met by his old friend, Admiral Doyle, the prospective Attack Force Commander. Smith later remembered that Doyle "was not very happy about the whole affair." They proceeded to Doyle's command ship, USS *Mount McKinley* (AGC 7). Smith's orders were to report his division directly to Commander in Chief, Far East, for operational control. His appointment with General MacArthur was set for 1730 that evening at the Dai Ichi building. Colonel Alpha L. Bowser, Jr., the division G-3, who had come out with the first echelon of Smith's staff, gave him a hurried briefing on the tentative plans for the division. "For the first time I learned that the division was to land at Inchon on 15 September," Smith wrote later.

On arriving at GHQ comfortably before the appointed time of 1730, Smith found that he was to meet first with Almond, who kept him waiting until 1900. Almond called most soldiers and officers "son," but when 58-year-old Almond addressed 57-year-old Smith as "son," it infuriated Smith. Almond further aggravated Smith by dismissing the difficulties of an amphibious operation as being "purely mechanical."

Having had his say, Almond ushered Smith into MacArthur's office. MacArthur, in a cordial and expansive mood, confidently told Smith that the 1st Marine Division would win the war by its landing at Inchon. The North Koreans had committed all their troops against the Pusan Perimeter, and he did not expect heavy opposition at Inchon. The operation would be somewhat "helter-skelter," but it would be successful. It was MacArthur's feeling that all hands would be home for Christmas, if not to the United States, at least to Japan.

Smith reported to Doyle his conviction that MacArthur was firm in his decision to land at Inchon on 15 September. Doyle replied that he thought there was still a chance to substitute Posung-Myun, a few miles to the south of Inchon, as a more likely landing site. Doyle was having his underwater demolition teams reconnoiter those beaches.

Next day, 23 August, Smith met again with Almond, this time accompanied by General Barr, commander of the 7th Infantry Division. When Smith raised the possibility of Posung-Myun as a landing site, Almond brushed him off, saying that any landing at Posung-Myun would be no more than a subsidiary landing.

Critical 23 August Conference Convenes

Smith was not invited to the 23 August conference. Nor was Shepherd. The all-important summit conference began with brief opening remarks by MacArthur. General Wright then outlined the basic plan which called for an assault landing by the 1st Marine Division directly into the port of Inchon. After the capture of Inchon, the division was to advance and seize, as rapidly as possible, Kimpo Airfield, the town of Yongdung-po, and the south bank of the Han River. The division was then to cross the river, capture Seoul, and seize the dominant ground to the north. Meanwhile, the 7th Infantry Division was to land behind the Marines, advance on the right flank, secure the south bank of the Han southeast of Seoul and the high ground north of Suwon. Thereafter, X Corps—1st Marine and 7th Infantry Divisions— would form the anvil against which the Eighth Army, breaking out of the Pusan Perimeter, would deliver the hammer blows that would destroy the North Korean Army.

After Wright's briefing, Doyle, as the prospective Attack Force commander, gave a thorough analysis of the naval aspects of the landing. Of greatest concern to Doyle were the tides. A point of contention was the length of the naval gunfire preparation. Doyle argued for three to four days of pre-landing bombardment by air and naval gunfire, particularly to take out the shore batteries. MacArthur's staff disputed this on the basis of the loss of tactical surprise. Admiral Sherman was asked his opinion and replied, "I wouldn't hesitate to take a ship up there."

"Spoken like a Farragut," said MacArthur.

With his concerns brushed aside, Doyle concluded his briefing with "the best that I can say is that Inchon is not impossible."

Collins questioned the ability of the Eighth Army to link up quickly with X Corps. He suggested Kunsan, to the south, as an alternate landing site. Sherman, in general terms, supported Collins' reservations. General MacArthur sat silently, puffing his pipe, for several moments. He then spoke and all agree that his exposition was brilliant. He dazzled and possibly confused his audience with an analogy from the French and Indian War, Wolfe's victory at Quebec: "Like Montcalm, the North Koreans will regard the Inchon landing as impossible. Like Wolfe I [can] take them by surprise."

As he himself remembered his summation years later in his memoirs:

> The Navy's objections as to tides, hydrography, terrain, and physical handicaps are indeed substantial and pertinent. But they are not insuperable. My confidence in the

MajGen David G. Barr, left, Commanding General of the U. S. Army's 7th Infantry Division meets with MajGen Edward M. Almond, Commanding General, X Corps, to discuss the Inchon landing. The 7th Division would land behind the Marines, advance on their right flank, and seize the commanding ground south of Seoul. National Archives Photo (USA) 111-SC349013

Navy is complete, and in fact I seem to have more confidence in the Navy than the Navy has in itself . . . . As to the proposal for a landing at Kunsan, it would indeed eliminate many of the hazards of Inchon, but it would be largely ineffective and indecisive. It would be an attempted envelopment which would not envelop. It would not sever or destroy the enemy's supply lines or distribution center, and would therefore serve little purpose. It would be a "short envelopment," and nothing in war is more futile. But seizure of Inchon and Seoul will cut the enemy's supply line and seal off the entire southern peninsula . . . . This in turn will paralyze the fighting power of the troops that now face Walker . . . . If my estimate is inaccurate and should I run into a defense with which I cannot cope, I will be there personally and will immediately withdraw our forces before they are committed to a bloody setback. The only loss then will be my professional reputation. But Inchon will not fail. Inchon will succeed. And it will save 100,000 lives.

Others at the conference recalled MacArthur's closing words at the conference as being: "*We* shall land at Inchon, and *I* shall crush them." This said, MacArthur knocked the ashes of his pipe out into a glass ashtray, making it ring, and stalked majestically out of the room.

General Collins still harbored reservations. He thought a main point had been missed: what was the strength of the enemy at Inchon and what was his capability to concentrate there?

Admiral Sherman was momentarily carried away by MacArthur's oratory, but once removed from MacArthur's personal magnetism he too had second thoughts. Next morning, 24 August, he gathered together in Admiral Joy's office the principal Navy and Marine Corps commanders. Present, in addition to Sherman and Joy, were Admirals Radford and Doyle and Generals Shepherd and Smith. Despite general indignation over MacArthur's failure to give due weight to naval considerations, it was now abundantly clear that the landing would have to made at or near Inchon. But perhaps there was still room for argument for another landing site with fewer hydrographic problems. Shepherd announced that he was going to see MacArthur once again before returning to Pearl Harbor and that he would make a

Command Structure for Inchon

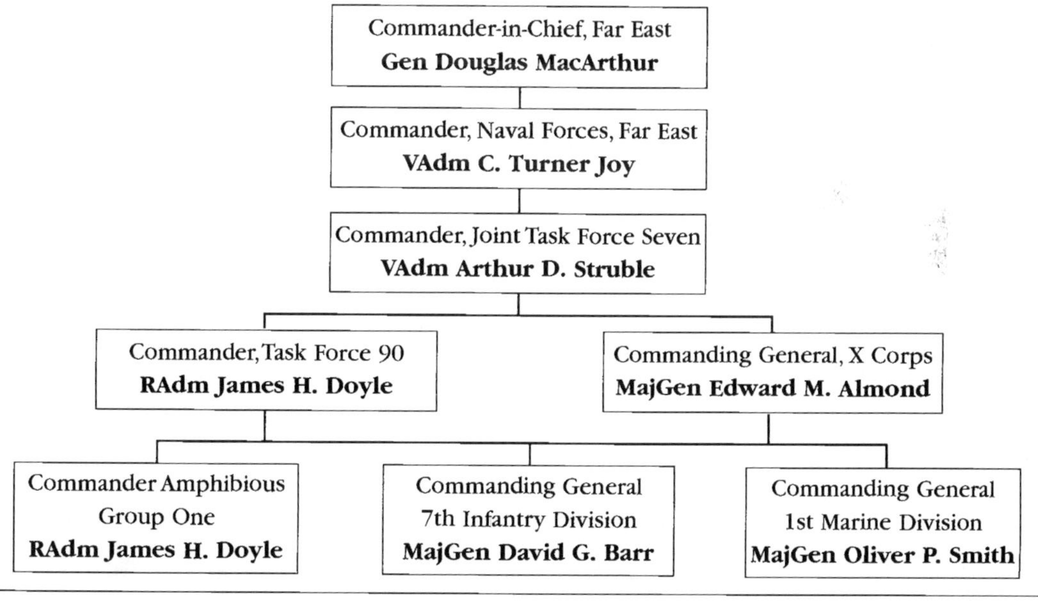

final plea for a landing south of Inchon in the vicinity of Posung-Myun.

Disappointment for General Shepherd

Shepherd, accompanied by Krulak, arrived at GHQ for his scheduled visit with MacArthur but was short-stopped by Almond who dismissed the Posung-Myun site, saying that Inchon had been decided upon and that was where the landing would be. The discussion became heated. Fortunately, MacArthur entered the room and waved Shepherd and Krulak into his office.

Shepherd had some expectation of being named the landing force commander. Admiral Sherman had recommended, without any great amount of enthusiasm, that Shepherd command X Corps for the operation because of his great amphibious experience and the expertise of his Fleet Marine Force, Pacific staff. General Wright on MacArthur's staff also recommended it, but a rumor was prevalent that Almond would get X Corps. MacArthur confirmed this intention, saying he would liked to have had Shepherd as commander, but that he had promised it to Almond. He asked if Shepherd would go along as his amphibious advisor. Shepherd hedged slightly. He said he would gladly go along as an observer.

Shepherd showed no rancor, then or later, at not getting command. He and Almond were both Virginians and both had gone to Virginia Military Institute—Almond, class of 1915 and Shepherd, class of 1917. Their personal relations were good but not close. Shepherd later characterized Almond as "an excellent corps commander. He was energetic, forceful, brave, and in many ways did a good job under difficult conditions." O. P. Smith would not come to share Shepherd's good opinion of Almond.

Plans Progress

The day following the 23 August conference, General Stratemeyer directed his staff to develop a FEAF plan to support the landing. The plan was to be separate from the CinCFE plan and was to provide mission direction for all combat aircraft not essential to the close support of the Eighth Army.

MacArthur, on 26 August, formally announced Almond's assignment as commanding general of X Corps. MacArthur had told him that he would continue, at the same time, to be the chief of staff of Far East Command. MacArthur's prediction was that Almond would soon be able to return to Tokyo. The landing at Inchon and subsequent capture of Seoul would end the war.

General Bradley's assessment of Almond was less than enthusiastic:

> Ned Almond had never commanded a corps—or troops in an amphibious assault. However, he and his staff, mostly recruited from MacArthur's headquarters, were ably backstopped by the expertise of the Navy and Marines, notably that of Oliver P. Smith, who commanded the 1st Marine Division, which would spearhead the assault.

MacArthur had not asked Collins and Sherman to approve his plan nor would they have had the authority to do so. The best they had to take back with them to Washington was a fairly clear concept of MacArthur's intended operations.

Collins and Sherman reported to Bradley and the other chiefs what they had learned about the Inchon plan, repeating their own misgivings. On 26 August, Bradley briefed President Truman and Secretary Johnson. The President was more optimistic than the chiefs.

'Conditional' Approval

On 28 August, the Joint Chiefs sent MacArthur a "conditional" approval, concurring in an amphibious turning movement, either at Inchon or across a favorable beach to the south. Chief "conditions" were that MacArthur was to provide amplifying details and keep them abreast of any modification of his plans. The Joint Chiefs specifically suggested preparation of an alternate plan for a landing at Kunsan.

X Corps dated its Operation Order No. 1, written largely by the facile pen of Colonel Forney, as 28 August; distribution was a day or so later. The 1st Marine Division "was charged with the responsibility as the Landing Force to assault INCHON, conduct beachhead operations, seize and protect KIMPO airfield, then advance to the HAN River line west of SEOUL. This achieved, the Division was further directed to seize SEOUL, and the commanding ground north of SEOUL, on order."

O. P. Smith's division staff, then on the *Mount McKinley*, was at half strength. Part of the remainder was enroute from the United States; part was with Craig's 1st Marine Brigade in the south of Korea. The brigade, although an organic part of the division, was still under the operational control of General Walker. Smith's staff, directed by Colonel Gregon A. Williams as chief of staff, worked well with Doyle's PhibGruOne staff. Above this harmonious relationship, the exact status of the more senior commands was indistinct and vaguely defined. From amidst a welter of paper, misunderstanding, ragged tempers, and sleep deprivation, Division Order 2-50, expanding on the corps order, emerged on 4 September.

Smith wrote later in the *Marine Corps Gazette*:

> By dedicated work on the part of the Division staff, with the wholehearted support of Adm Doyle's PhibGruOne staff, within three days a detailed plan for the Inchon Landing was drawn up, and two days later an advance planning draft of 1stMarDiv OpO 2-50 (Inchon Landing) was issued.

Time available for planning was so short that the assault regiments, contrary to amphibious doctrine, would get rigid landing plans drawn up completely by division.

The always dapper General Stratemeyer, seeking to solidify his contention that he was General MacArthur's tactical air commander, conferred with Joy, Struble, and Almond at CinCFE headquarters on 30 August. All that he could get was a general agreement on the adequacy of a CinCFE 8 July directive, "Coordination of Air Effort of Far East Air Forces and United States Naval Forces, Far East." Building on that, Stratemeyer sent a message to MacArthur, the gist of it being: "It is recognized that ComNavFE must have control of air operations within the objective area during the amphibious phase. Air operations outside of the objective area are part of the

Junior officers and enlisted Marines did not get a briefing on their unit's role in the landing until embarked in amphibious shipping enroute to the objective area. However, by then, because of leakage to the press, it was an open secret that the Marines were going to land at Inchon.

Department of Defense Photo (USMC) A2681

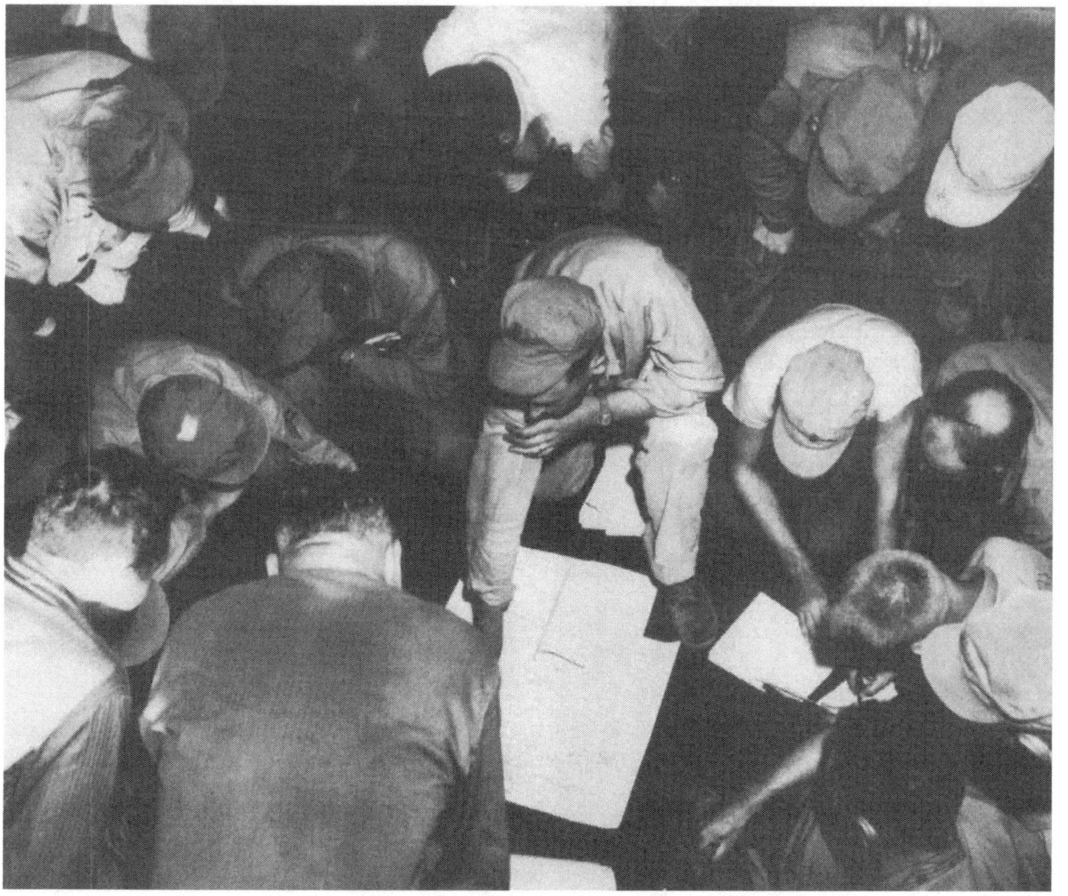

overall air campaign, and during the amphibious phase contribute to the success of the amphibious operation."

MacArthur's headquarters issued Operation Order No. 1 on 30 August, but neither a copy of this order nor any other amplifying detail had reached Washington by 5 September. On that date the chiefs sent a further request for details to MacArthur. Choosing to consider the 28 August JCS message to be sufficient approval, MacArthur dismissed the request with a brief message, stating "the general outline of the plan remains as described to you."

Later he would write that his plan "was opposed by powerful military in Washington." He knew that Omar Bradley, the JCS chairman, had recently testified to Congress that large-scale amphibious operations were obsolete. He disliked Bradley personally and derisively referred to him as a "farmer."

Both Bradley and Truman came from Missouri working-class families and were proud of it. A routine had been established under which the Joint Chiefs kept Truman informed, usually by a personal briefing by Bradley, of the current situation in Korea.

On 7 September, MacArthur received a JCS message which he said chilled him to the marrow of his bones. The message asked for an "estimate as to the feasibility and chance of success of projected operation if initiated on planned schedule."

The offending message reminded MacArthur that all reserves in the Far East had been committed to the Eighth Army and all available general reserves in the United States—except for the 82d Airborne Division—had been committed to the Far East Command. No further reinforcement was in prospect for at least four months. In light of this situation, a fresh evaluation of Inchon was requested.

MacArthur Protests

An indignant MacArthur fired back an answer, the concluding paragraph of which said: "The embarkation of the troops and the preliminary air and naval preparations are proceeding according to schedule. I repeat that I and all my commanders and staff officers are enthusiastic for and confident of the success of the enveloping movement."

The last sentence was manifestly not true. Lack of enthusiasm was readily apparent at all levels of command.

Next day, 8 September, the JCS sent MacArthur a short, contrite message: "We approve your plan and the President has been informed." The phrase "the President has been informed" annoyed MacArthur. To him it implied something less than presidential approval and he interpreted it as a threat on President Truman's part to overrule the Joint Chiefs. General Collins, for one, had no recollection of Truman ever expressing any doubt about the success of the Inchon landing or any inclination to override the actions of the JCS with respect to the operation.

Beach Reconnaissance

According to the intelligence available to General Smith, the enemy had about 2,500 troops in the Inchon-Kimpo region, including at least two battalions of the *226th Independent Marine Regiment* and two companies of the *918th Artillery Regiment*. The North Koreans had apparently prepared strong defensive positions. Reconnaissance reports indicated 106 hard targets, such as gun emplacements, along the Inchon beaches.

Some of the best beach intelligence was obtained by Navy offshore reconnaissance. Best known are the exploits of Lieutenant Eugene F. Clark, ex-enlisted man and an experienced amphibious sailor. He and two South Koreans left Sasebo on 31 August on board the British destroyer HMS *Charity*, transferred the next morning to a South Korean frigate, and landed that evening on Yong-hong-do, 14 miles off Inchon and one of the hundreds of islands that dotted Korea's west coast. The islanders were friendly. Clark organized the island's teenagers into coastwatching parties and commandeered the island's only motorized sampan. For two weeks he fought a nocturnal war, capturing more sampans, sending agents into Inchon, and testing the mud flats for himself. His greatest accomplishment was discovering that one of the main navigation lights for Flying Fish Channel was still operable. GHQ at Tokyo instructed him to turn on the light at midnight on 14 September. This he would do.

Anticipated hydrographic conditions were much more frightening than the quality of expected enemy resistance. Doyle's Attack Force would have to thread its way from the Yellow Sea through the tortuous Flying Fish Channel. As had already been determined, the 15th of September was the best day of the month because of the height and spacing of the tides. The morning high tide—an incredible 31.5 feet—would be at 0659 and the evening high tide at 1919. In between these times, as the tide fell, the currents would rip out of the channel at seven or eight knots, exposing mud flats across which even amphibian tractors

Terrain Handbook No. 65: Seoul and Vicinity (GHQ, Far East Command, 16 August 1950)

This pre-landing aerial photograph shows clearly the convoluted nature of the Inchon "beachhead." MajGen Oliver P. Smith, commanding the landing force, considered Wolmi-do, the island at the lower left of the photo, the key to the whole situation. Seizure of Wolmi-do would precede the main landings on Inchon itself.

could not be expected to crawl.

Wolmi-do: Key to Operation

Wolmi-do ("Moon Tip Island"), the long narrow island that formed the northern arm of Inchon's inner harbor, was thought to have about 500 defenders. Wolmi-do harbor was connected to the Inchon dock area by a 600-yard-long causeway. "Wolmi-do," wrote Smith, was "the key to the whole operation."

Brigade staff officers, headed by their chief of staff, Colonel Edward W. Snedeker, were called to Japan from Pusan. They recommended that the 3d Battalion, 5th Marines, be used for the assault of Wolmi-do.

Smith's plan, as it emerged, was to take Wolmi-do on the morning tide by landing the 3d Battalion, 5th Marines, across Green Beach. Then would come a long wait of 12 hours until the evening tide came in and the remainder of the division could continue the landing. The rest of the 5th Marines would cross Red Beach to the north of Wolmi-do, while Puller's 1st Marines landed over Blue Beach in the inner harbor to the south. Designation of the landing sites as "beaches" was misleading; the harbor was edged with cut-granite sea walls that would have to be scaled or penetrated.

Colonel Snedeker recommended that the new 1st Korean Marine Corps Regiment be added to the troop list. The assignment of the Republic of Korea (ROK) Marines to the division was approved by GHQ on 3 September. The Eighth Army was instructed to provide them weapons.

Almond asked Smith to take part in a war-gaming of the operation. Smith saw it as nothing more than a "CPX" or command post exercise and a waste of precious time. He sent a major in his place.

Almond inspected units of Barr's 7th Division at their camps—Fuji, McNair, McGill, Drake, and Whittington—between 31 August through 3 September. His aide,

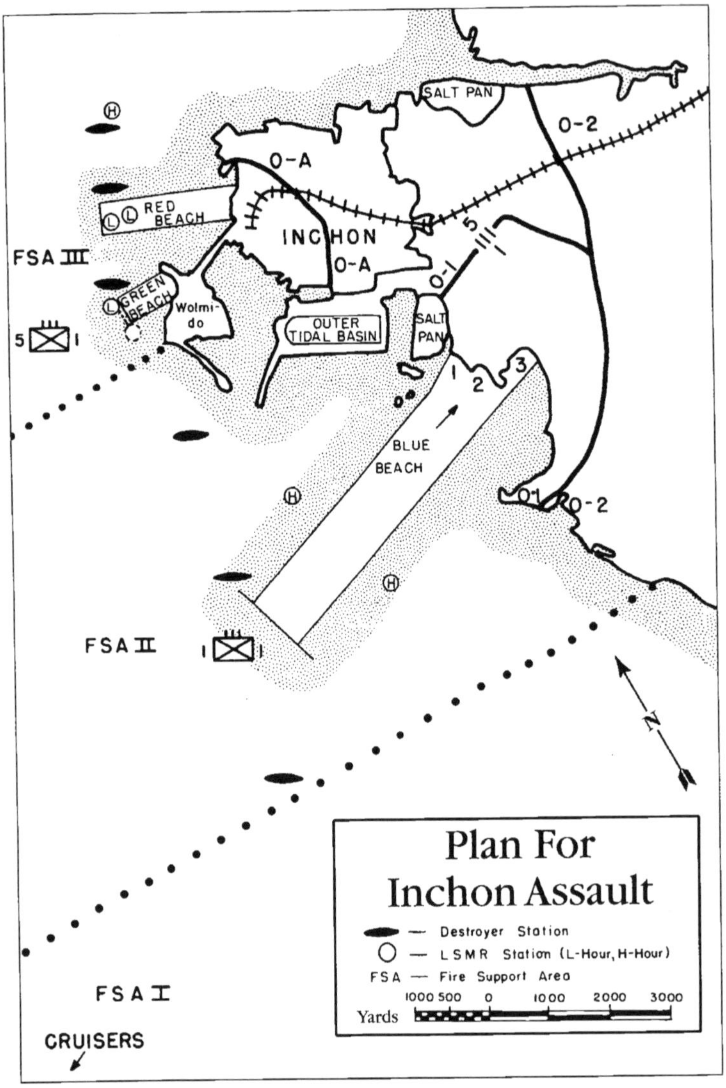

First Lieutenant Haig, accompanied him and took extensive notes. With few exceptions, Almond gained a "good" to "excellent" impression of the units he visited.

On the morning of 2 September Almond met with the officers of his Corps staff who were involved in his war game. He pointed out the necessity for frequent visits to subordinate units by commanding officers and the need for strong, well-organized, defenses for Corps headquarters. "The front line is the perimeter of the place where you happen to be," said Almond.

Meanwhile, the main body of the 1st Marine Division arrived at Kobe, Japan—except for the 5th Marines, which was still at Pusan, and the 7th Marines, which was still at sea.

Typhoon Jane Disrupts Embarkation

Typhoon Jane, with winds up to 74 miles an hour, struck Kobe on 3 September. Two feet of water covered the docks. One ship, with all the division's signal gear, settled to the bottom at her pier. All unloading and loading stopped for 24 hours. Property sergeants, called in from the outlying battalions, worked frantically to sort out their units' gear.

Adding to General Smith's worries, the availability of the 5th Marines was now challenged. General Walker, deeply involved in the bitter defense of the Naktong Bulge, strongly opposed the release of this now-seasoned regiment from his Eighth Army. To meet Walker's objections, and influenced by his own favorable impression of the 7th Division, Almond sent Colonel Forney, now the Marine Deputy Chief of Staff, X Corps, to ask O. P. Smith whether the 7th Marines would arrive in time to be substituted for the 5th Marines, or alternatively, if not, would the 32d Infantry be acceptable?

A conference on the proposed substitution was held on the evening of 3 September. Present, among others, were Generals Almond and Smith and Admirals Joy, Struble, and Doyle. Strangely, General Barr, the 7th Division's commander, was not there. The discussion became heated. Smith argued that the proposal went beyond a considered risk. If the substitution were made, he declared, he would change his scheme of maneuver. He would call off the landing of the 1st Marines over Blue Beach and give them the 5th Marines' mission of landing on Red Beach with the 32d Infantry following behind.

Admiral Struble (Shepherd thought him "slippery") resolved the contretemps by suggesting that a regiment of Barr's 7th Division be immediately embarked to stand off Pusan as a floating reserve, allowing the release of the 1st Provisional Marine Brigade. In General Smith's mind, Almond's proposal exemplified the wide gulf separating Army and Marine Corps thinking. As Colonel Bowser, General Smith's operations officer, remembered it, Doyle and Smith "came back about 11 o'clock having won their point, that the [Marine] brigade must come out of the Pusan perimeter and be part of our landing force."

The *Mount McKinley*, flagship of the Attack Force—with Smith on board so as to be in a better position to supervise the out-loading—set sail from Tokyo for Kobe on 4 September, arriving there early the next afternoon. That evening Smith called a conference of all available Marine Corps commanders to stress the urgency of the operation.

Almond Inspects Marines

A day later, 6 September, General Almond came to Kobe to inspect 1st Marine Division units. He lunched with the staff noncommissioned officers at Camp Otsu accompanied by General Smith and Lieutenant Colonel Allan Sutter, then visited the 2d and 3d Battalions of the 1st Marines. Afterwards he went to Camp Sakai near Osaka to see the 11th Marines, the division's artillery regiment commanded by Colonel James H. Brower, and was favorably impressed." He commented in his diary: "A large percentage of the troops were drawn from active Marine reserve units . . . . The Army should have done likewise but did not."

President Harry S. Truman and Marine Commandant Gen Clifton B. Cates exchange warm greetings at a Marine Corps field demonstration at Quantico in June 1950, 10 days before the outbreak of the Korean War. This friendly relationship dissolved when Truman, in an ill-advised note, called the Marine Corps "the Navy's police force."

Department of Defense Photo (USMC) A407260

In the evening Smith and his staff briefed him on the division's operation plan. Again Almond was favorably impressed, but he thought Smith's planned subsequent moves ashore too slow and deliberate. He stressed to Smith the need for speed in capturing Kimpo Airfield and Seoul itself. Smith was less impressed with Almond, saying: "The inspection consisted [of Almond] primarily questioning men, I suppose for the purpose of finding out what made Marines tick."

In the 1st Marine Division, operational planning trickled down to the battalion level. The 3d Battalion, 1st Marines, under Lieutenant Colonel Thomas L. Ridge, had steamed comfortably to Japan in the *General Simon B. Buckner* (AP 123) and was ensconced in what had been the barracks for a battalion of the 24th Infantry Division at Otsu on the south shore of Lake Biwa. There was no room for field training and the best the battalion could do was road-bound conditioning marches. The commanding officer and the three majors in the battalion were summoned to a meeting on board the regimental command ship berthed in Kobe. There had been a plethora of rumors, but now for the first time they learned officially that they were to land at Inchon. The regimental S-2, Captain Stone W. Quillian, went over the beach defenses, tapping a large map studded with suspected weapons emplacements. The S-3, Major Robert E. Lorigan, then briefed the scheme of maneuver. The 3d Battalion would be the right flank unit of the main landing. These were the D-Day objectives. *Tap, tap.* This piece of high ground was the battalion's objective. *Tap, tap.* This hook of land on the extreme right flank had to be taken. *Tap, tap.* The landing would be at 1730; it would be dark at 1900. There were no enthusiastic cheers from the listeners.

Then the regimental commander, Chesty Puller, got to his feet. "You people are lucky," he growled. "We used to have to wait every 10 or 15 years for a war. You get one every five years. You people have been living by the sword. By God, you better be prepared to die by the sword."

The troop list for the landing force totalled 29,731 persons, to be loaded out in six embarkation groups. Four groups would load out of Kobe, one group out of Pusan, and one group—made up of the Army's 2d Engineer Special Brigade—out of Yokohama. Not all units could be combat loaded; some compromises had to be accepted.

One Marine Corps unit that was not ready to go was the 1st Armored Amphibian Tractor Battalion, activated but not yet combat ready. The Army's Company A, 56th Amphibian Tractor Battalion, was substituted.

President Writes Letter

As the Marines combat loaded their amphibious ships at Kobe, the Pacific edition of *Stars and Stripes* reached them with a story that President Truman had called them "the Navy's police force." This compounded a previously perceived insult when the President labeled the United Nations intervention in Korea a "police action." The enraged Marines chalked on the tarpaulins covering their trucks and tanks, "Horrible Harry's Police Force" and similar epithets.

What had happened was that on 21 August, Congressman Gordon L. McDonough of California had written President Truman a well-intentioned letter urging that the Marines be given a voice on the Joint Chiefs of Staff. The President fired back a feisty note: "For your information the Marine Corps is the Navy's police force and as long as I am President that is what it will remain. They have a propaganda machine that is almost equal to Stalin's . . . . The Chief of Naval Operations is the Chief of Staff of the Navy of which the Marines are a part."

He had dictated the letter to his secretary, Rose Conway, and sent it without any member of his staff seeing it.

McDonough inserted the letter into the Congressional Record where it appeared on 1 September. The story reached the newspapers four days later and a great public outcry went up. By five o'clock the next afternoon Truman's advisors had prevailed upon him to send an apology to General Cates: "I sincerely regret the unfortunate choice of language which I used." Truman, in further fence-mending, in company with Cates, made a surprise visit two mornings later at a Marine Corps League convention coincidentally being held in Washington's Statler Hotel and charmed his audience.

Pulling Together the Landing Force

General Craig's 1st Provisional Marine Brigade was relieved of its combat commitment at midnight, 5 September. The brigade had done most of its fighting with a peacetime structure, that is, at about two-thirds its authorized wartime strength: two rifle companies to a battalion instead of three, four guns to an artillery battery instead of six. The 5th Marines did not get a third company for its three infantry battalions until just before mounting out for Inchon.

The Korean 1st Marine Regiment, some 3,000 men, commanded by Lieutenant Colonel Kim Sung Eun, arrived in Pusan on 5 September to join the 1st Marine Division. They were in khaki uniforms including cloth caps, and equipped with Japanese rifles and machine guns. The South Korean Marines were issued American uniforms—including helmets—and each was given one day on the rifle range to fire his new American weapons.

Built around a cadre drawn from the ROK Navy, the Korean Marine Corps ("KMCs" to the U.S. Marines) had been activated 15 April 1949. Company-size units had first deployed to southern Korea, and then to Cheju Island, to rout out Communist-bent guerrillas. After the North Korean invasion, the KMCs, growing to regimental size, had made small-scale hit-and-run raids along the west coast against the flank of the invaders.

Craig assigned Lieutenant Colonel Charles W. Harrison, until recently the executive officer of the 6th Marines at Camp Lejeune, as liaison officer to the KMCs. His party, given a radio jeep, was made up of three corporal radiomen, and a corporal driver. Harrison was well-chosen. His parents had been missionaries in Korea. He himself had graduated from the foreign high school in Pyongyang in 1928 and he had a working knowledge of Korean.

While the 5th Marines were

loading out, a paper, marked "Confidential" and giving specifics on a landing beach at Kaesong, was widely distributed and one or more copies were purposely "lost." Perhaps the word got back to the North Koreans.

The amphibious assault transport *Henrico* (APA 45) known to the fleet as "Happy Hank," had brought the 1st Battalion, 5th Marines, to Pusan. Now the ship received the same battalion, its numbers, thinned by the fighting in the Pusan Perimeter, now brought up to war strength. The Navy crew did their best to provide a little extra for their Marine passengers. The wardroom was made available to the officers 24 hours a day.

Marguerite "Maggie" Higgins, a movie-star-pretty blonde reporting on the war for the New York *Herald-Tribune* occupied one of the few staterooms. She had been a war correspondent in Europe during the last years of World War II and had been in Korea since the beginning of the new war. Ribald rumors as to her imagined nocturnal associations inevitably circulated throughout the ship.

Major General Field Harris, Commanding General, 1st Marine Aircraft Wing—O. P. Smith's aviator counterpart—arrived in Tokyo on 3 September. His forward echelon of the 1st Marine Aircraft Wing, was informed of the Inchon-Seoul operation three days later. Planning for the employment of Marine air was completed on 9 September. Marine Aircraft Group 33, relieved of its close support role in the Pusan Perimeter, would be the operating element. Harris and his forward echelon embarked at Kobe on 10 September as Tactical Air Command, X Corps.

Meanwhile, Almond continued his restless visits and inspections. On 9 September, General Barr briefed him on the 7th Infantry Division's plan of operations. Almond thought the plan adequate, but was concerned over possible problems of liaison and coordination with the 1st Marine Division. Events would prove him right

Almond's Good Ideas

A restive General Almond formed, for commando work, a Special Operations Company, X Corps, sometimes called a "Raider Group," under command of Colonel Louis B. Ely, Jr., USA. With Almond's encouragement, Ely proposed a raid to seize Kimpo Airfield. Almond asked Smith for 100 Marine volunteers to join the Special Operations Company; Smith, skeptical of the mission and unimpressed by Ely, stalled in providing Marines and the request was cancelled. As it turned out, Ely and his company would make an approach to the beach, but the distance from ship to shore proved too great for rubber boats.

Brigadier General Henry I. Hodes, USA, the assistant division commander of the 7th Infantry Division, visited Smith on the *Mount McKinley* on 9 September. Almond, still concerned by Smith's deliberate manner, had come up with yet another idea for the swift seizure of Kimpo. Almond's new plan called for landing a battalion of the 32d Infantry on Wolmi-do the evening of D-Day. It would "barrel" down the road to Seoul in trucks and tanks provided by the Marines. Smith, horrified by a plan he considered tactically impossible, told Hodes that he had no tanks to lend him.

The Secretary of the Navy, alerted by parents' complaints that underage sons were being sent to Korea, on 8 September sent a last-minute order to remove Marines under 18 before sailing, reducing the landing force by about 500 men. Those who were close to being 18 were held in Japan on other duties and eventually found their way to the division as replacements.

Second Typhoon

Weathermen said that a second typhoon, "Kezia," was following close behind "Jane." Rear Admiral Arleigh A. Burke, USN, had arrived in Tokyo from Washington to be Admiral Joy's deputy chief of staff. Burke attempted to make an office call on MacArthur to express his concerns regarding the coming typhoon and was blocked by Almond. Burke refused to discuss the matter with Almond and went back to his office. By the time he got there, a message was waiting that MacArthur would see him. Burke hurried back to GHQ and explained to MacArthur that if the typhoon came up and blew west there could be no landing on the 15th or 16th.

"What do we do, Admiral?" asked MacArthur.

"We sail early," said Burke. MacArthur agreed.

Navy meteorologists had first picked up signs of Kezia off the Mariana Islands on 6 September. Whipping up winds of 100 miles per hour, the typhoon moved steadily toward Japan and the East China Sea. Most endangered were the amphibious ships of Admiral Doyle's Attack Force. The route for all six transport groups to Inchon placed them squarely in the path of the on-coming oriental hurricane.

Both Doyle and O.P. Smith, the two who would bear the burden of directing the actual landing, were painfully aware that all the normal steps of preparing for an amphibious operation were either

Photo courtesy of LtCol Leo J. Ihli, USMC (Ret)

Marines prime an F4U-4B of VMF 323 for take-off from the deck of the light aircraft carrier Badoeng Strait *(CVE 116) standing off Korea. VMF-214, embarked in sister carrier* Sicily *(CVE 118), played a companion role in close support of the assault. The bent-wing Corsairs would prove once again to be ideal close support aircraft.*

being compressed or ignored completely in order to squeeze the operation into an impossibly short time frame. During World War II, at least three months would have been spent in planning and training for an operation of this magnitude. Beginning with Guadalcanal, a rehearsal—or rehearsals—was considered essential. For Inchon there would be no rehearsal. Doyle wryly concluded that a good deal would depend upon how skillfully the individual coxswains could perform in finding their way to the beaches.

Captain Martin J. "Stormy" Sexton, a World War II Raider and now aide-de-camp to General Smith, said later: "There was not even time for landing exercises by the LVTs. Some of the LVT crews had not even had the opportunity to try their engines out in the water and paddle around."

Execution

Marine aircraft squadrons VMF-214 and VMF-323 began the softening-up of Wolmi-do on 10 September with the delivery of napalm. Operating from the decks of the light carriers *Sicily* (CVE 118) and *Badoeng Strait* (CVE 116) ("Bing-Ding" to the Marines and sailors), the Marine fliers burned out most of the buildings on the island. Strikes by Navy aircraft from the big carriers *Valley Forge* (CV 45), *Philippine Sea* (CV 47), and *Boxer* (CV 21) continued for the next two days.

Joint Task Force 7 (JTF 7) was officially activated under Admiral Struble the following day, 11 September. Almond and X Corps would be subordinate to Struble and JTF 7 until Almond assumed command ashore and JTF 7 was dissolved.

Preliminary and diversionary air and naval gunfire strikes were roughly divided into 30 percent delivered north of Inchon, 30 percent south, and 40 percent against Inchon itself. Except for a few gunnery ships held back to protect the flanks of the Pusan Perimeter, JTF 7—in its other guise, the Seventh

Joint Task Force Seven

VAdm Arthur D. Struble

Task Force 90	Attack Force	RAdm James H. Doyle
Task Force 91	Blockade and Covering Force	RAdm Sir William G. Andrewes
Task Force 92	X Corps	MajGen Edward M. Almond
Task Force 99	Control and Reconnaissance Force	RAdm George R. Henderson
Task Force 77	Fast Carrier Group	RAdm Edward C. Ewen
Task Force 79	Service Squadron	Capt Bernard L. Austin
ROK Naval Forces		Cdr Michael L. Luosey[1]

[1] Liaison and Advisor

Fleet—included all the combatant ships in the Far East. Among them were three fast carriers, two escort carriers, and a British light carrier. In the final count, the force numbered some 230 ships, including 34 Japanese vessels, mostly ex-U.S. Navy LSTs (landing ships, tank) with Japanese crews. The French contributed one tropical frigate, *La Grandiere*, which arrived at Sasebo with a five-month supply of wine and a pin-up picture of Esther Williams, but no coding machine.

Mount McKinley, with Doyle, Smith, and their staffs on board, got underway from Kobe the morning of 11 September—a day ahead of schedule because of the approach of Typhoon Kezia—and steamed for Sasebo. Winds of the typhoon whipped up to 125 miles per hour. Doyle was gambling that Kezia would veer off to the north.

Almond held a last meeting at GHQ on 12 September to deal with the urgency for an early sailing because of the threat of Kezia. General Shepherd, General Wright, and Admiral Burke attended. That afternoon General MacArthur and his party left Haneda airport to fly to Itazuke air base. From there they would go by automobile to Sasebo.

MacArthur Goes to Sea

Because of the storm the *Mount McKinley* was late in reaching port. MacArthur's party waited in the Bachelor Officers Quarters, passing the time having sandwiches. It was close to midnight before the *Mount McKinley* rounded the southern tip of Kyushu and docked at Sasebo. MacArthur and his party boarded the ship and she was underway again within an hour. With General Shepherd came his G-3, Colonel Victor H. Krulak, and his aide and future son-in-law, Major James B. Ord, Jr.

MacArthur had five generals in his party—Shepherd, Almond, and Wright, and two others: Major General Courtney Whitney—his deputy chief of staff for civil affairs, but more importantly his press officer—and Major General Alonzo P. Fox. Fox was chief of staff to MacArthur in his capacity as "SCAP" (Supreme Commander Allied Powers) and Lieutenant Haig's father-in law. Absent from the group was Lieutenant General George Stratemeyer, USAF, who had had some expectation of accompanying MacArthur as his air boss. In assignment of spaces, MacArthur grandly ignored traditional ship protocol and took over Doyle's cabin. Doyle moved to his sea cabin off the flag bridge. Almond appropriated the ship's captain's cabin. O. P. Smith managed to keep his stateroom.

After breakfast on the morning of the 13th, Admiral Doyle led the embarked flag officers in a tour of the *Mount McKinley*, hoping to impress the Army generals that amphibious operations required specialization. MacArthur did not go along.

The absence of General Stratemeyer from MacArthur's party was a clear signal that the Navy had been successful in keeping the Air Force from operating within the amphibious objective area—a circle with a 100-mile radius drawn around Inchon. There would be no FEAF operations within this radius unless specifically requested by Struble. MacArthur remained above these doctrinal squabbles.

Operation 'Common Knowledge'

Neither General MacArthur nor Admiral Struble favored extensive air and naval gunfire preparation of the objective area, primarily because it would cause a loss of tactical surprise. Their concern was largely academic. All sorts of leakage circulated in Japan—and even reached the media in the United States—that an amphibious operation was being mounted out with a probable target of Inchon. At the Tokyo Press Club the impending landing was derisively called "Operation Common Knowledge." The North Korean command almost certainly heard these rumors and almost equally certain had tide tables for Inchon. Mao Tse Tung is supposed to have pointed at Inchon on a map of Korea and have said, "The Americans will land here."

American intelligence knew that the Russians had supplied mines, but how many had been sown in Flying Fish Channel? The lack of time and sufficient minesweepers made orderly mine-sweeping operations impossible.

'Sitting Ducks'

The pre-landing naval gunfire bombardment began at 0700 on 13 September with a column of cruisers and destroyers coming up the channel. The weather was good, the sea calm. Four cruisers—*Toledo* (CA 133), *Rochester* (CA 124), HMS *Kenya*, and HMS *Jamaica*—found their bombardment stations several miles south of Inchon and dropped anchor. Six destroyers—*Mansfield* (DD 728), *DeHaven* (DD 727), *Lyman K. Swenson* (DD 729), *Collett* (DD 730), *Gurke* (DD 783), and *Henderson* (DD 785)—continued on past the cruisers and were about to earn for themselves the rueful title of "Sitting Ducks."

What appeared to be a string of mines was sighted in the vicinity of Palmi-do. The destroyers opened fire with their 40mm guns and the mines began to explode. Leaving the *Henderson* behind to continue shooting at the mines, the five other destroyers steamed closer to their objectives. *Gurke* anchored 800 yards off Wolmi-do, which was being pounded by carrier air.

The remaining four destroyers took station behind *Gurke*. Just before 1300 they opened fire. Within minutes return fire came blazing back from hidden shore batteries. *Collett* took five hits, knocking out her fire direction system; her guns switched to individual control. *Gurke* took two light hits. *DeHaven* was slightly damaged. *Lyman K. Swenson* felt a near miss that caused two casualties. After an hour's bombardment the destroyers withdrew. One man had been killed—ironically Lieutenant (Junior Grade) David Swenson, nephew of the admiral for whom the destroyer was named—and eight were wounded.

From their more distant anchorage, the cruisers picked up the bombardment with 6-inch and 8-inch salvos. After that the carrier aircraft resumed their attack.

Next day, 14 September, five of the destroyers came back (the damaged *Collett* was left behind) and banged away again. At first the destroyers drew feeble return fire. By the time they withdrew 75 minutes later, having delivered 1,700 5-inch shells, there was no return fire at all. The Navy, with considerable satisfaction, reported Wolmi-do now ready for capture.

Attack Force Gathers

Admiral Doyle had won his gamble against the typhoon. The Yellow Sea was quiet and all elements of the Attack Force were in place off Inchon. General Craig's embarked 1st Provisional Marine Brigade, having arrived from Pusan, was formally dissolved on 13 September and its parts returned to the control of the parent division. Craig became the assistant division commander.

The Attack Force eased its way up Flying Fish Channel so as to be in the transport area before daylight on 15 September. General MacArthur spent a restless night. Standing at the rail of the *Mount McKinley* in the darkness, he entertained certain morbid thoughts, at least as he remembered them later in his *Reminiscences*: "Within five hours 40,000 men would act boldly, in the hope that 100,000 others manning the thin defense lines in South Korea would not die. I alone was responsible for tomorrow, and if I failed, the dreadful results would rest on judgment day against my soul."

George Gilman, an ensign in the *Mount McKinley*, had less lofty thoughts: "None of us boat group officers had ever had any experience operating under such tidal conditions before, let alone ever having been involved in an amphibious landing . . . . As the morning of September 15 approached, we realized we had all the ingredients for a disaster on our hands."

Destination Wolmi-do

L-hour was to be 0630. At 0545, the pre-landing shore bombardment began. Lieutenant Colonel Robert D. "Tap" Taplett's 3d Battalion, 5th Marines, was boated by 0600. The carrier-based Marine Corsairs completed their last sweep of the beach 15 minutes later.

"G Company was to land to the right of Green Beach in the assault, wheel right, and seize the domi-

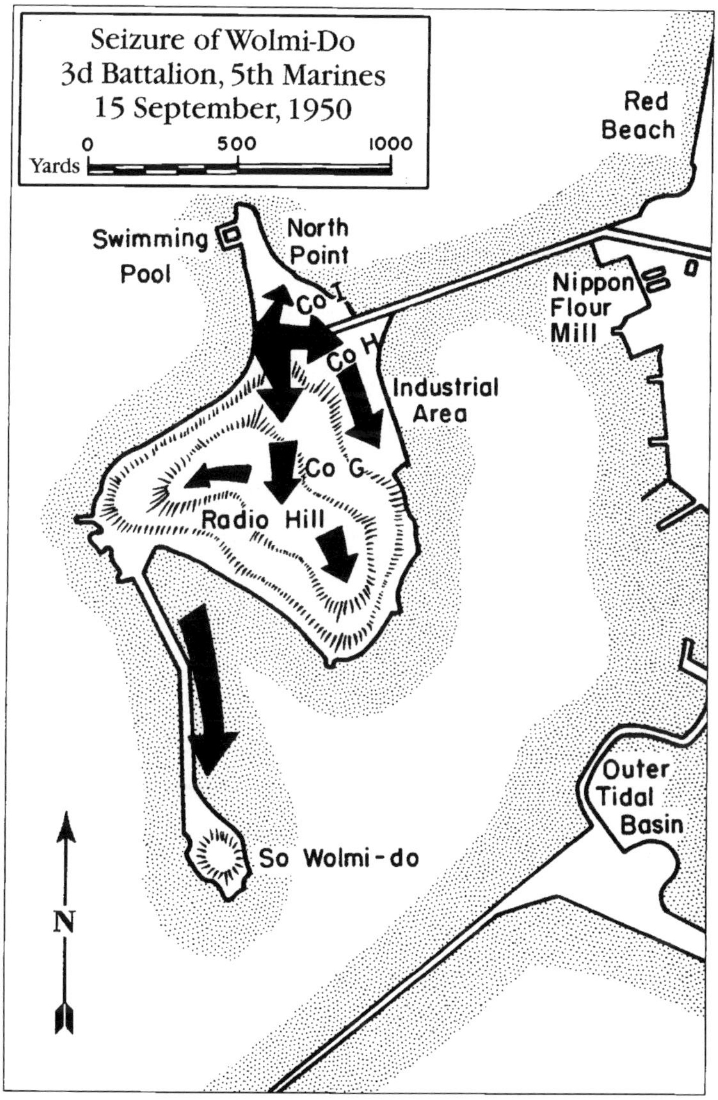

nant hill mass on the island, Radio Hill," remembered Robert D. "Dewey" Bohn (then a first lieutenant; he would retire a major general). His company was embarked in the fast destroyer transport *Diachenko* (APD 123). She stopped her engines at about 0300, the troop compartment lights came on, and reveille sounded over the public address system.

Most of the Marines were already awake. They hoped for the traditional "steak and eggs" pre-landing breakfast of World War II; instead they got scrambled powdered eggs, dry toast, and canned apricots. At about first light, Company G went over the side and down the cargo nets into the bobbing LCVPs, which then cleared the ship and began to circle.

Three LSMRs—medium landing ships converted to rocket ships—sent their loads of thousands of 5-inch rockets screeching shoreward toward Wolmi-do. The island seemed to explode under the impact. Then the landing craft began the run to Green Beach. MacArthur, Shepherd, Almond, Smith, Whitney, and Doyle all watched from the flag bridge of the *Mount McKinley*.

Seven LCVPs brought in the first wave, one platoon of Company G on the right and three platoons of Company H on the left. The landing craft converged on the narrow beach—scarcely 50 yards wide—and grounded at 0633, three minutes behind schedule. The remainder of the two assault companies came in as the second wave two minutes later. Resistance was limited to a few scattered shots.

Captain Patrick E. Wildman, commanding Company H, left a small detachment to clear North Point and then plunged across the island toward his objectives—the northern nose of Radio Hill and the shoreline of the burning industrial area facing Inchon. After a short pause to reorganize, Bohn took Company G towards the southern half of Radio Hill, 105 meters high. Resistance was half-hearted. At 0655, Sergeant Alvin E. Smith, guide of the 3d Platoon, secured an American flag to the trunk of a shattered tree. MacArthur, watching the action ashore from his swivel chair on the bridge of the *Mount McKinley*, saw the flag go up and said, "That's it. Let's get a cup of coffee."

Ten tanks—six M-26 Pershings and four modified M-4A3 Shermans, all under Second Lieutenant Granville G. Sweet—landed in the third wave at 0646 from three utility landing ships (LSUs). They crunched their way inland, poised to help the infantry.

Lieutenant Colonel Taplett land-

Department of Defense Photo (USMC) A2686

Reveille in the amphibious ships went at 0300 on the morning of 15 September. Marines hoped for the traditional "steak-and-eggs" D-day breakfast of World War II, but most transports fed simpler fare, such as powdered eggs and canned apricots. Breakfast on board the landing ships was even more spartan.

ed from his free boat a few minutes later. At almost the same time, Captain Robert A. McMullen brought in the fourth wave bearing Company I, the battalion reserve. His company, following behind Company H, encountered an angry nest of about a platoon of by-passed North Koreans. A flurry of hand grenades was exchanged. McMullen signaled Sweet's tanks to come forward. A Sherman with a dozer blade sealed the die-hard North Koreans in their holes.

Moving on to the near end of the causeway that stretched to Inchon itself, McMullen found more North Korean defenders hiding in a cave. One of Sweet's tanks fired a 90mm round into the mouth of the cave. There was a muffled explosion and 30 dazed and deafened North Koreans came staggering out with their hands above their heads. "Captured forty-five prisoners . . . meeting light resistance," radioed Taplett at 0745 to the *Mount McKinley*.

Wildman's Marines were finding it slow going in the ruins of the industrial area. Taplett ordered Bohn to take the rest of Radio Hill and by 0800 the high ground was Marine Corps property.

'Wolmi-do Secured'

Once again Taplett radioed the *Mount McKinley*, this time: "Wolmi-do secured."

With the success of the Marine landing blaring over the loud-speakers, MacArthur left the bridge

By 0655, the 3d Battalion, 5th Marines, had landed on Wolmi-do and had an American flag flying at the top of a shell-blasted tree. An hour later the battalion commander reported resistance as light and 45 dazed prisoners taken.

National Archives Photo (USMC) 127-GK-2341-A2694

Department of Defense Photo (USMC) A2798

Some North Korean defenders of Wolmi-do stubbornly remained in their cave-like positions and had to be burned out by flamethrowers. U.S. Marines were readily distinguishable at this stage of the war by their wear of camouflage helmet covers and leggings.

to pen a message to Admiral Struble in his flagship *Rochester*: "The Navy and Marines have never shone more brightly than this morning."

Ashore, Taplett consolidated his gains. His three rifle companies, by prearranged plan, took up defensive positions facing Inchon. The empty swimming pool at the tip of North Point became a stockade for prisoners.

At about 10 o'clock Taplett ordered Bohn to take Sowolmi-do, an islet dangling to the south of Wolmi-do with a lighthouse at the end of the causeway. Bohn sent Second Lieutenant John D. Counselman, leader of his 3d Platoon, with a rifle squad and a section of tanks. As a prelude to the assault, a flight of Corsairs drenched Sowolmi-do with napalm. Covered by the two tanks and a curtain of 81mm mortar fire, Counselman's riflemen crossed the narrow causeway, taking fire from a hill honey-combed with emplacements. Flamethrowers and 3.5-inch rocket launchers burned and blasted the dug-in enemy. Seventeen were killed, 19 surrendered, and eight or more managed to hide out. The lighthouse was taken and the job completed in less than two hours. Three Marines were wounded, bringing Taplett's casualties for the day to none killed, 17 wounded.

Word was passed that some of the North Koreans who had escaped were trying to swim for Inchon. A number of Bohn's Marines lined up rifle-range fashion and shot at what they saw as heads bobbing in the water. Others dismissed the targets as imaginary. Mopping up of the island was completed by noon.

Taplett, growing restless and seeing no sign of enemy activity, proposed to division that he make an assault on the city from his present position or at least a reconnaissance in force. Smith responded to his proposal with a firm negative.

Waiting for Evening Tide

The remainder of the division was steaming toward the inner transport area. There would now be a long wait until the evening tide swept in and the assault regi-

M-26 Pershing tanks, new to the Marines, began to land in the third wave at Wolmi-do and were soon put to use against North Korean fortified positions. A tank-infantry patrol assaulted and took Sowolmi-do, an islet dangling at the end of a causeway from the main island.

Department of Defense Photo (USMC)

Department of Defense Photo (USMC) A2723

Marines from the 3d Battalion, 5th Marines, escorted a steady stream of prisoners back to Green Beach on the seaward side of Wolmi-do. Landing ships and craft could beach as long as the tide was high, but once the tide receded they would be left high and dry on the mud flats.

ments could be landed. Marines, standing at the rail of their transports, strained their eyes looking for their intended beaches but could see nothing but smoke. The bombardment, alternating between naval gunfire and air strikes, continued.

During the course of the afternoon, Admiral Struble had his admiral's barge lowered into the water from the *Rochester* ("Roach-Catcher"). He swung by the *Mount McKinley* to pick up General MacArthur for a personal reconnaissance from close offshore of Wolmi-do and the harbor. Almond and Shepherd went with them.

They swung close to the seawall fronting the harbor. "General," said Shepherd, "You're getting in mighty close to the beach. They're shooting at us." MacArthur ignored the caution.

Naval gunfire and carrier air sought to hit everything that could be found in the way of targets within a 25-mile radius of Inchon. (The D-Day action for the aircraft on board the carrier *Boxer* was labeled "Event 15" and consisted of a strike with 12 F4U Corsairs and five AD Skyraiders.) The smoke of the bombardment and from burning buildings mixed with the rain so that a gray-green pall hung over the city.

H-Hour for the main landing was 1730. Lieutenant Colonel Raymond L. Murray's 5th Marines, minus the 3d Battalion already ashore on Wolmi-do, was to land over Red Beach, to the left and north of Wolmi-do. Murray's regiment was to seize the O-A line, a blue arc on the overlay to the division's attack order. On the ground O-A line swung 3,000 yards from Cemetery Hill on the north or left flank, through Observatory Hill in the center, and then through a maze of buildings, including the British Consulate,

A corpsman bandages the forearm of a wounded North Korean prisoner on Wolmi-do. He and other prisoners were moved to one of the several prison stockades that were set up on the landing beaches.

Department of Defense Photo (USMC) A2802

Department of Defense Photo (USA) SC348839

Gen MacArthur indulged his passion for visiting the "front." During the interval between the morning and evening landings he personally "reconnoitered" the Inchon beaches in VAdm Arthur D. Struble's barge. Struble sits to MacArthur's right. On his left is Army MajGen Courtney Whitney, often-called MacArthur's "press secretary."

until it reached the inner tidal basin.

The 1st and 2d Battalions, 5th Marines, under Lieutenant Colonels George R. Newton and Harold S. Roise respectively, would land abreast across Red Beach. The new 1st ROK Marine Regiment would follow them ashore.

Newton and Roise had the Pusan Perimeter behind them, but not much other infantry experience. Newton, commissioned in 1938 from the Naval Academy, was with the Embassy Guard at Peking when World War II came on 7 December 1941 and spent the war as a prisoner of the Japanese. Roise, commissioned from the University of Idaho in 1939, had served at sea during the war.

In the assault, Newton's 1st Battalion and Roise's 2d Battalion would come away from the attack transports *Henrico* and *Cavalier* (APA 37) in landing craft. Both battalions would land in column of companies across the seawall onto narrow Red Beach. Newton, on the left, was to take Cemetery Hill and the northern half of Observatory Hill. Roise, on the right, was to take his half of Observatory Hill, the British Consulate, and the inner tidal basin.

"Two things scared me to death," said Roise of the landing plan. "One, we were not landing on a beach; we were landing against a seawall. Each LCVP had two ladders, which would be used to climb up and over the wall. This was risky . . . . Two, the landing was scheduled for 5:30 p.m. This would give us only about two hours of daylight to clear the city and set up for the night."

Captain Francis I. "Ike" Fenton, Jr., commander of Company B in Newton's battalion, sharply

Lieutenant Colonel Raymond L. Murray

Seldom does a Marine Corps regiment go into combat with a lesser grade than full colonel in command. But when Brigadier General Edward Craig arrived at Camp Pendleton in July 1950 to form the 1st Provisional Marine Brigade for service in Korea he found no reason to supplant the commanding officer of the 5th Marines, Lieutenant Colonel Raymond Murray. The tall, rangy Texan was an exception to the general rule. He had already made his reputation as a fighter and of being a step ahead of his grade in his assignments. As a major at Guadalcanal he had commanded the 2d Battalion, 6th Marines, and for his conspicuous gallantry had earned his first Silver Star.

After Guadalcanal, came Tarawa for the battalion and a second Silver Star for Murray, now a lieutenant colonel. Finally, at Saipan, although he was painfully wounded, Murray's control of his battalion was such that it brought him a Navy Cross.

Novelist Leon Uris served in Murray's battalion. Later, when he wrote his book *Battle Cry*, he used Murray as his model for "High Pockets" Huxley, his hard-charging fictional battalion commander.

Born in Alhambra, California, in 1913, Murray grew up in Harlingen, Texas. When he accepted his commission in July 1935, after graduating from Texas A&M College, then the incubator of many Army and Marine officers, he had behind him four years of the Army's Reserve Officers Training Corps and two years of the Texas National Guard. He had also starred at football and basketball. After attending Basic School, then in the Philadelphia Navy Yard, he was detailed to the 2d Marine Brigade in San Diego. The brigade went to troubled China a year later. Murray served for a short time in Shanghai, then moved to a prized slot in the Embassy Guard in Peking. He came back to San Diego in 1940 and returned to the 2d Marine Brigade which within months expanded into the 2d Marine Division. A 1st Provisional Marine Brigade was pulled out of the 2d Division in the summer of 1941 for service in Iceland. Murray, now a captain and soon to be a major, went with it. He was back in San Diego in April 1942 and in October sailed with the 6th Marines for the war in the Pacific.

He came home in August 1944 and served at Quantico, Camp Lejeune, Hawaii, and Camp Pendleton. Promotions were slow after 1945 and Murray was still a lieutenant colonel when the Korean War began in 1950. As commander of the infantry element of the later-day 1st Provisional Marine Brigade in the "fire brigade" defense of the Pusan Perimeter, he received his third and fourth Silver Stars for his staunch leadership.

At Inchon, Major General O. P. Smith gave Murray and his now-seasoned regiment the more complicated northern half of the landing. After Inchon and Seoul, Murray would continue in command through the Chosin Reservoir campaign. That battle in subzero weather brought him the Army's Distinguished Service Cross as well as his second Navy Cross. Finally, in January 1951 he was promoted to colonel.

Coming home from Korea in April 1951, he attended the National War College and then was handpicked to command The Basic School, since World War II at Quantico. Next he served at Camp Pendleton and Camp Lejeune. A promotion to brigadier general came in June 1959. Assignments in Okinawa, then Pendleton again, and Parris Island followed. Serving at Headquarters Marine Corps in 1967 as a major general, he was ordered to Vietnam as Deputy Commander, III Marine Amphibious Force. His strong physique finally failed him. He was invalided home in February 1968 to Bethesda Naval Hospital where he remained until his retirement on 1 August 1968. He now lives in Oceanside, California, close to Camp Pendleton.

Department of Defense Photo (USMC) A42922

Department of Defense Photo (USMC) A2865

Navy transports stand off Inchon and Wolmi-do before the landing. Amphibious lift for Inchon, some of it literally borrowed back from the Japanese, was a rusty travesty of the great amphibious armadas of World War II.

remembered the characteristics of Red Beach:

> Once on the beach there was an open area of about 200 yards. The left flank was marked by Cemetery Hill. From the sea it looked like a sheer cliff. To the right of Cemetery Hill was a brewery, some work shops, and a cotton mill. Further to the right and about 600 yards in from the beach was Observatory Hill, overlooking the entire landing area and considered critical; it was the regimental objective. Further to the right was a five-story office building built of concrete and reinforced steel.

Captain John R. Stevens' Company A was to land on the right flank. In the assault would be the 2d Platoon under Second Lieutenant Francis W. Muetzel and the 1st Platoon under Gunnery Sergeant Orval F. McMullen. In reserve was the 3d Platoon under First Lieutenant Baldomero Lopez, who had joined the company as it loaded out from Pusan.

Three miles to the south of the 5th Marines, Chesty Puller's 1st Marines was to land across Blue Beach. Puller's mission was to secure the O-1 line, a 4,000-yard arc that went inland as deep as 3,000 yards, and then hooked around to the left to cut off Inchon from Seoul.

Blue Beach One, 500 yards wide, had its left flank marked by a salt evaporator. What looked to be a road formed the boundary to the south with Blue Beach Two.

The 2d Battalion, 1st Marines, was under affable, white-haired Lieutenant Colonel Allan Sutter. After landing over Blue Beach One, he was to take a critical road junction about 1,000 yards northeast of the beach, and Hill 117, nearly two miles inland, which commanded Inchon's "back door" and the highway to Seoul, 22 miles away.

Sutter, a graduate of Valley Forge Military Academy and Dartmouth College, had gained his Marine Corps commission in 1937 through the Platoon Leaders Course, a program under which college students spent two summers at Quantico to qualify as second lieutenants. He then spent a year at the Basic School in Philadelphia before being assigned troop duties. During World War II, Sutter was a signal officer at Guadalcanal, Guam, and Okinawa.

Blue Beach Two, also 500 yards wide, had its left flank marked by the supposed road and its right flank by a narrow ramp jutting seaward. A cove, further to the right, named at the last minute "Blue Beach Three," offered an alternate or supplementary landing site. Ridge, with the 3d Battalion, was to cross the seawall girdling Blue Beach Two and take Hill 233, a mile southeast of the beach, and, on the extreme right, a small cape, flanking Blue Beach and topped by Hill 94.

At best, the four assault battalions coming across Red and Blue Beaches would have but two hours of high tide and daylight to turn the plan into reality. Smith, after fully committing his two regiments, would have nothing left as a division reserve except two half-trained Korean Marine battalions.

Assaulting Red Beach

It would be a long ride to Red Beach for the 1st and 2d Battalions of the 5th Marines. Troops began debarking from the transports at about 1530. "As you climb down that net into the LCVP you're scared," remembered Private First Class Doug Koch of Company D, 5th Marines. "What keeps you

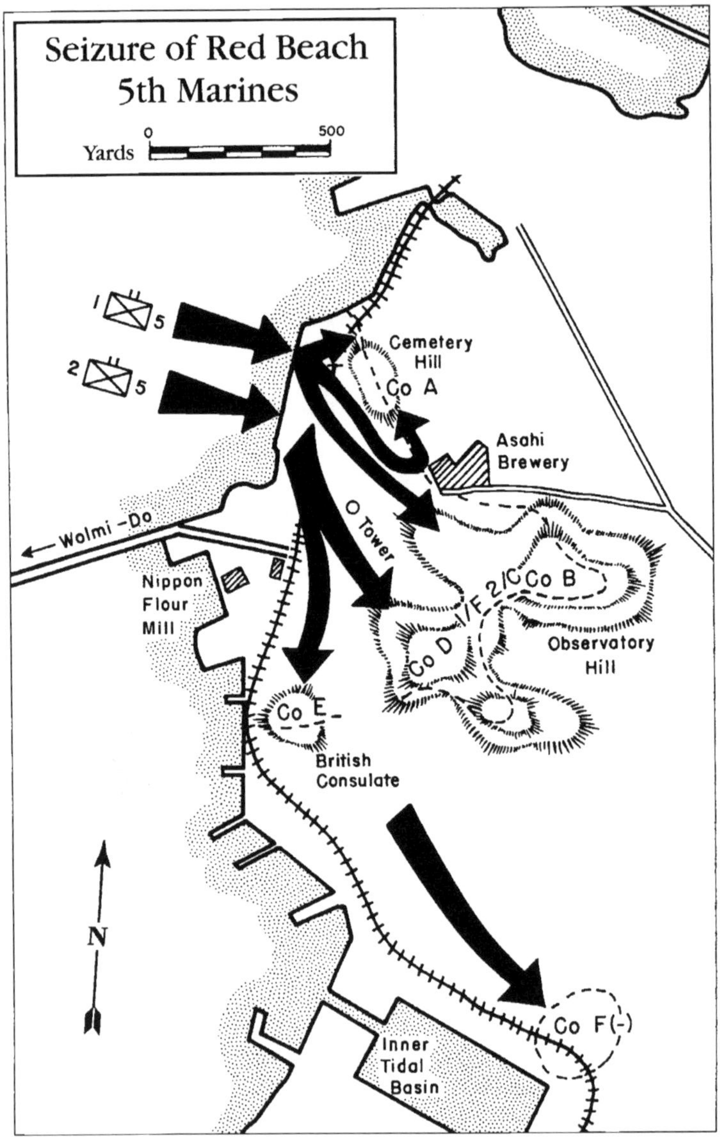

going is knowing this is what you have to do."

The *Horace A. Bass* (APD 124), the Red Beach control vessel, slowly steamed ahead with a long file of landing craft "trailing behind like a brood of ducklings."

The supporting rocket ships let go with a final fusillade of some 6,500 5-inch rockets. The resulting cloud of dust and smoke completely masked the beach area. The *Horace A. Bass*, an escort destroyer converted into a high-speed transport and anxious to get into the fight, banged away with her 5-inch guns. She then dipped her signal flag and the first wave headed for Red Beach.

The eight LCVPs in the first wave crossed the line of departure at H-8 with 2,200 yards to go. The four boats on the left carried the two assault platoons of Company A. Captain Steven's mission was to take Cemetery Hill and to secure the left flank of the beachhead. The four boats on the right carried the assault elements of Captain Samuel Jaskilka's Company E, which was to clear the right flank of the beach and then capture the hill that held the British Consulate.

As the first wave passed the mid-way point, two squadrons of Marine Corps Corsairs—VMF-214 under Lieutenant Colonel Walter E. Lischied and VMF-323 under Major Arnold A. Lund—came in to strafe both Red and Blue Beaches. They exhausted their loads and flew away. Not satisfied, Captain Stevens called for further air strikes against Red Beach. Four Navy A-4D Skyraiders made strafing passes until the wave had only 30 yards to go.

On the right, First Lieutenant Edwin A. Deptula's 1st Platoon, Company E, hit the seawall at 1731, one minute behind schedule. Designated Marines threw grenades up over the seawall, and after they exploded, Deptula took his platoon up the scaling ladders. A few stray rounds whined overhead.

Deptula pushed inland about 100 yards to the railroad tracks against no resistance. The rest of Company E landed about 10 minutes later. Captain Jaskilka (who would retire as a four-star general) quickly re-organized his company near the Nippon Flour Company building just south of the beachhead. Deptula's platoon continued down the railroad tracks to the British Consulate. Jaskilka sent another platoon to cross the railroad tracks and then move up the slope of 200-foot-high Observatory Hill.

On the left flank it was not quite that easy. One of the four landing

Gen Oliver P. Smith Collection, Marine Corps Research Center

Aerial photo of Red Beach shows the pounding it took in the pre-landing naval gunfire and air attacks. The 1st and 2d Battalions, 5th Marines, landed across this beach immediately north of the causeway leading to Wolmi-do.

craft, with half the 1st Platoon, Company A, on board, lagged behind with engine trouble. The remaining three boats reached the seawall at 1733. Sergeant Charles D. Allen took his half of the 1st Platoon over the wall and received fire from his north flank and from a bunker directly to his front. Several Marines went down.

To Allen's right, Second Lieutenant Frank Muetzel found a breach in the seawall and brought his 2d Platoon ashore. Facing them was a pillbox. Two Marines threw grenades and six bloody North Korean soldiers came out. Cemetery Hill loomed ahead, but Muetzel's immediate objective was Asahi Brewery. He slipped south of Cemetery Hill and marched unopposed down a street to the brewery. There was a brief indulgence in green beer.

Sergeant Allen, with his half-platoon, was making no progress against the bunker to his front. The second wave landed, bringing in the 3d Platoon under Baldomero Lopez and the missing half of the 1st Platoon. Too many Marines were now crowded into too small a space.

Lopez charged forward alone. He took out the bunker with a grenade and moved forward against a second bunker, pulling the pin from another grenade. Before he could throw it, he was hit. The grenade dropped by his side. He smothered the explosion with his body. This gained him a posthumous Medal of Honor. Two Marines went against the bunker with flamethrowers. They were shot down but the bunker was taken.

Captain Stevens's boat landed him in Company E's zone of action. Unable to get to his own company, he radioed his executive officer, First Lieutenant Fred F. Eubanks, Jr., to take charge. Stevens then radioed Muetzel to leave the brewery and get back to the beach where he could help out.

On the way back, Muetzel found a route up the southern slope of Cemetery Hill and launched an assault. The summit was alive with North Koreans, but there was no fight left in them. Dazed and spiritless from the pounding they had taken from the air and sea, they threw up their hands and surrendered. Muetzel sent them down to the base of the hill under guard.

Eubanks' Company E Marines meanwhile had bested the obstructing bunker with grenades and a flamethrower. His 1st and 2d Platoons pushed through and joined Muetzel's 2d Platoon. At 1755, 25 minutes after H-Hour, Captain Stevens fired an amber flare, signaling that Cemetery Hill was secure. It had cost his company eight Marines killed and 28 wounded.

Coming in on the third and fourth waves, Company C, 1st Battalion, was to take the northern half of Observatory Hill, and Company D, 2d Battalion, was to take the southern half. It did not work out quite that way. Parts of Companies C and D were landed on the wrong beaches. Company C, once ashore, had to wait 12 minutes for its commander, Captain Poul F. Pedersen. In Pedersen's boat was the fifth wave commander who had decided to tow a stalled LCVP. Once ashore, Pedersen had trouble sorting out his company from amongst the jumble of Marines that had gathered in the center of the beach.

Maggie Higgins, the *Herald-Tribune* correspondent, came off the *Henrico* in Wave 5 along with John Davies of the Newark *Daily News*, Lionel Crane of the London *Daily Press*, and a photographer. As their landing craft hit the seawall, the wave commander, First Lieutenant Richard J. "Spike" Schening, urged on his Marines with, "Come on you big, brave Marines. Let's get the hell out of here."

The photographer decided he had had enough and that he would go back to the *Henrico*. Maggie considered doing the same, but then, juggling her typewriter, she, along with Davies and Crane, followed Schening over the seawall.

Eight LSTs crossed the line of departure, as scheduled, at 1830 and were headed for the seawall. Seeing the congestion on Red Beach, the skippers of the LSTs concluded that the Marines were held up and could not advance.

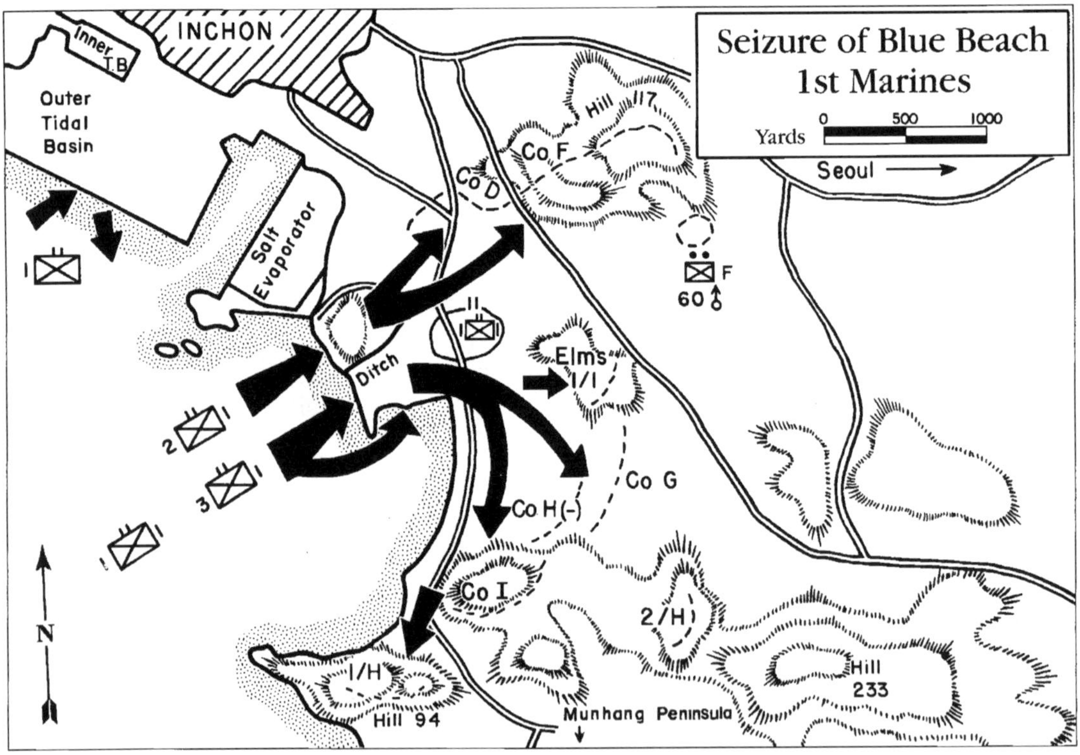

Department of Defense Photo (USMC) A3701

Marines enroute to Red Beach go over the side of their assault transport, down the cargo net hand-over-hand, and into the waiting LCVP, a version of the famous "Higgins boat" of World War II.

The lead LST received some mortar and machine-gun fire and fired back with its own 20mm and 40mm guns. Two other LSTs joined in. Unfortunately, they were spraying ground already occupied by the Marines.

The LST fire showered Muetzel's platoon, holding the crest of Cemetery Hill. Muetzel pulled back his platoon. As his Marines slid down the hill, they came under fire from a North Korean machine gun in a building on Observatory Hill. A chance 40mm shell from one of the LSTs knocked out the gun. Weapons Company and Headquarters and Service Company of Roise's 2d Battalion landed about 1830 and came under LST fire that killed one Marine and wounded 23 others.

By 1900, all eight LSTs had stopped firing and were nestled against the seawall. By then Second Lieutenant Byron L. Magness had taken his 2d Platoon, Company C, reinforced by Second Lieutenant Max A. Merritt's 60mm mortar section, up to the saddle that divided the crest of Observatory Hill. Their radios were not working and they had no flares. They had to inform the beach of their success by sending back a runner.

Meanwhile, the 1st Battalion's reserve company—Company B under Captain "Ike" Fenton—had landed in the 2d Battalion's zone. Lieutenant Colonel Newton ordered Fenton to assume Company C's mission and take the northern half of Observatory Hill. Six Marines were wounded along the way, but by about 2000 Fenton was at the top and tied in with the Magness-Merritt platoon.

In the right half of the regimental zone of action, Roise was getting the congestion on the beach straightened out. Company D, commanded by First Lieutenant H. J. Smith, had followed Company E ashore, but had landed to the left in the 1st Battalion zone. Smith (called "Hog Jaw" to make up for his non-existent first and second names) understood that Jaskilka's Company E was already on the crest of Observatory Hill. Under that assumption he started his company in route column up the street leading to the top of the hill. An enemy machine gun interrupted his march. After a brisk firefight that caused several Marine casualties, the enemy was driven off and Company D began to dig in for the night. A platoon from Company F, the battalion reserve, filled in the gap between Company D and the Magness-Merritt positions. The only part of the O-A line that was not now under control was the extreme right flank where the line ended at the inner tidal basin.

Maggie Higgins, after seeing the war, such as it was, found a boat

Department of Defense Photo (USMC) A3190

Marines go over the seawall forming the sharp edge of Red Beach. The Marine on the ladder has been identified as 1stLt Baldomero Lopez. Moments later he would give his life and earn a posthumous Medal of Honor.

on Red Beach that was returning to the *Mount McKinley*, where, after the personal intercession of Admiral Doyle, she was allowed to stay for the night. She slept on a stretcher in the sick bay. Next day, Admiral Doyle specified that in the future women would be allowed on board only between the hours of nine in the morning and nine at night. (About a month later, Maggie's transportation orders were modified. She would still be allowed on board any Navy ship but would have to be chaperoned by a female nurse.)

Murray, the regimental commander, came ashore at about 1830 and set up his command post at the end of the causeway that led from the mainland to Wolmi-do. Roise wished to stay where he was for the night, but Murray ordered him to reach the tidal basin. Company F, under Captain Uel D. Peters, faced around in the dark and plunged forward. Shortly after midnight, Roise reported that his half of the O-A line was complete.

Assaulting Blue Beach

The confusion was greater on Blue Beach than on Red Beach. Amphibian tractors, rather than landing craft, were used for the assault. The seawall was in disrepair with numerous breaks up which it was presumed the amphibian tractors could crawl. The 18 Army armored amphibians (LVT[A]s) forming the first wave crossed the line of departure at 1645 and headed toward Inchon. At four knots they needed three-quarters of an hour to hit the beach at H-Hour.

The soldiers had the compasses and seamanship to pierce the smoke and reached the beach on time. The second and following

First Lieutenant Baldomero Lopez

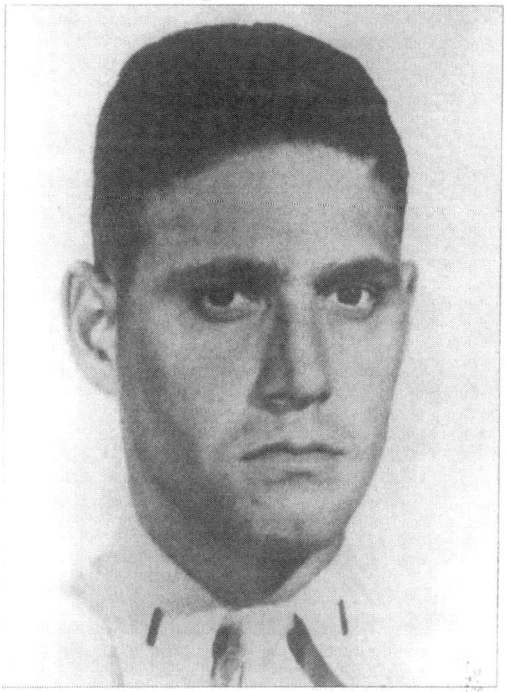

Department of Defense Photo (USMC) A43985

Baldomero Lopez was always eager. During World War II, he was 17 when he enlisted in the Navy in July 1943. Most thought him a Mexican American, but his father, also named Baldomero, as a young man had come to Tampa from the Asturias region of Spain. Los Asturianos, the men of Asturias, are known for their valor and honor.

He was appointed from the fleet to the Naval Academy in July 1944. His class, 1948A, was hurried through in three years. Lucky Bag, his class book, called him "one of the biggest hearted, best natured fellows in the brigade." Otherwise he does not seem to have been exceptional. His nickname at the Academy was "Lobo." This changed to "Punchy" after he came into the Marine Corps in June 1947, because it was generally believed that he had boxed while at Annapolis. After Basic School he stayed on at Quantico as a platoon commander in the Platoon Leaders Class. In 1948, he went to North China as part of a Marine presence that was in its last days. He served first as a mortar section leader and then as a rifle platoon commander at Tsingtao and Shanghai.

When the Marines closed out in China, he came back to Camp Pendleton. In the early summer of 1950, when the formation of the 1st Provisional Marine Brigade stripped the 1st Marine Division dry, he asked to be included but was left behind. He went out, however, to Korea in the draft that was sent to Pusan to fill the 5th Marines to war strength before embarking for Inchon. He was given the 3d Platoon, Company A, 1st Battalion.

Secretary of the Navy Dan Kimball presented the posthumous Medal of Honor to his father and mother at ceremonies in Washington on 30 August 1951.

Citation:

> For conspicuous gallantry and intrepidity at the risk of his life above and beyond the call of duty as a Rifle Platoon Commander of Company A, First Battalion, Fifth Marines, First Marine Division (Reinforced), in action against enemy aggressor forces during the Inchon invasion in Korea on 15 September 1950. With his platoon, First Lieutenant Lopez was engaged in the reduction of immediate enemy beach defenses after landing with the assault waves. Exposing himself to hostile fire, he moved forward alongside a bunker and prepared to throw a hand grenade into the next pillbox whose fire was pinning down that sector of the beach. Taken under fire by an enemy automatic weapon and hit in the right shoulder and chest as he lifted his arm to throw, he fell backward and dropped the deadly missile. After a moment, he turned and dragged his body forward in an effort to retrieve the grenade and throw it. In critical condition from pain and loss of blood, and unable to grasp the hand grenade firmly enough to hurl it, he chose to sacrifice himself rather than endanger the lives of his men and, with a sweeping motion of his wounded right arm, cradled the grenade under him and absorbed the full impact of the explosion. His exceptional courage, fortitude and devotion to duty reflect the highest credit upon First Lieutenant Lopez and the United States Naval Service. He gallantly gave his life for his country.

waves did not do so well. Rain and smoke had completely blotted out any view of the beach. From the bridge of his ship, the Blue Beach control officer watched the first two or three waves disappear into the smoke. He requested permission to stop sending any further waves ashore until he could see what was happening to them. Permission was denied.

As Major Edwin H. Simmons, the commander of Weapons Company, 3d Battalion, 1st Marines, remembered it:

> We had been told that a wave guide would pick us up and lead us to the line of departure . . . . Two LCVPs did come alongside our wave.

Department of Defense Photo (USMC) A2816

A key objective for the 5th Marines was the 200-foot-high Observatory Hill. Both the 1st and 2d Battalions converged on the hill with Marines from Company B taking the weather station on its top.

The first was filled with photographers. The second was loaded with Korean interpreters. Two of these were hastily dumped into my LVT, apparently under the mistaken notion that I was a battalion commander. Both interpreters spoke Korean and Japanese; neither spoke English. Time was passing, and we were feeling faintly desperate when we came alongside the central control vessel. I asked the bridge for instructions. A naval officer with a bullhorn pointed out the direction of Blue Two, but nothing could be seen in that direction except mustard-colored haze and black smoke. We were on our way when our path crossed that of

Marines setup a temporary barricade on the causeway to Inchon, after mopping up and consolidating their positions on Wolmi-do. Although not expecting a counterattack, they position a 3.5-inch rocket launcher and a machine gun just in case. The 3.5-inch rocket launcher proved itself adequate against the vaunted T-34 tank.

National Archives Photo (USMC) 127-N-A2747

Department of Defense Photo (USMC)

Amphibian tractors (LVTs) churn away from the landing ships (LSTs) that brought them to Inchon. "Amtracks" were used chiefly for the assault of Blue Beach within the inner harbor.

another wave. I asked if they were headed for Blue Two. Their wave commander answered, "Hell, no. We're the 2d Battalion headed for Blue One." We then veered off to the right. I broke out my map and asked my LVT driver if he had a compass. He looked at his instrument panel and said, "Search me. Six weeks ago I was driving a truck in San Francisco."

The nine Army LVT(A)s making up the first wave for Blue Beach One got ashore on schedule, but found themselves boxed in by an earth slide that blocked the exit road. The remaining nine Army armored amphibian tractors, forming Wave 1 for Blue Beach Two, made it to the seawall shortly after H-Hour but were less successful in getting ashore. The "road" separating Blue One and Two turned out to be a muck-filled drainage ditch. After exchanging fire with scattered defenders in factory buildings behind the seawall, the Army vehicles backed off and milled around, getting intermixed with the incoming troop-carrying Waves 2 and 3.

From his seat on the bridge of the *Mount McKinley*, MacArthur, surrounded by his gaggle of generals and admirals, peered through the gathering gloom of smoke, rain, and darkness and listened to the reports crackling over the loudspeaker. From his perspective, all seemed to be going well.

Lieutenant Colonel Sutter's second wave landed elements of both his two assault companies, Company D, under Captain Welby W. Cronk, and Company F, under Captain Goodwin C. Groff, across Blue Beach One shortly after H-Hour. Some of his amphibian tractors hung up on a mud bank about 300 yards offshore and their occupants had to wade the rest of the way. Most of Sutter's last three waves, bringing in his reserve, Company E, drifted to the right. As Sutter reported it: "For some unknown reason the third, fourth, and fifth waves were diverted from landing either on Beach BLUE-1 or along the rock causeway by a control boat. Instead they were directed to the right of the two beaches prescribed for the regiment and landed at Beach BLUE-3."

Wave 2 for Blue Beach Two, with Ridge's assault companies, passed through the Army tractors, Company G under Captain George C. Westover on the left, Company I under First Lieutenant Joseph R. "Bull" Fisher on the right. They reached the seawall about 10 minutes after H-Hour. The tractors bearing Company G formed up in column and muddled their way up the drainage ditch. Company I went over the seawall using aluminum ladders, some of which buckled. Assault engineers from Captain Lester G. Harmon's Company C, 1st Engineer Battalion, reached the wall and rigged cargo nets to help the later waves climb ashore.

Ridge, the 3d Battalion commander, accompanied by his executive office, Major Reginald R. Myers, seeing the congestion on Blue

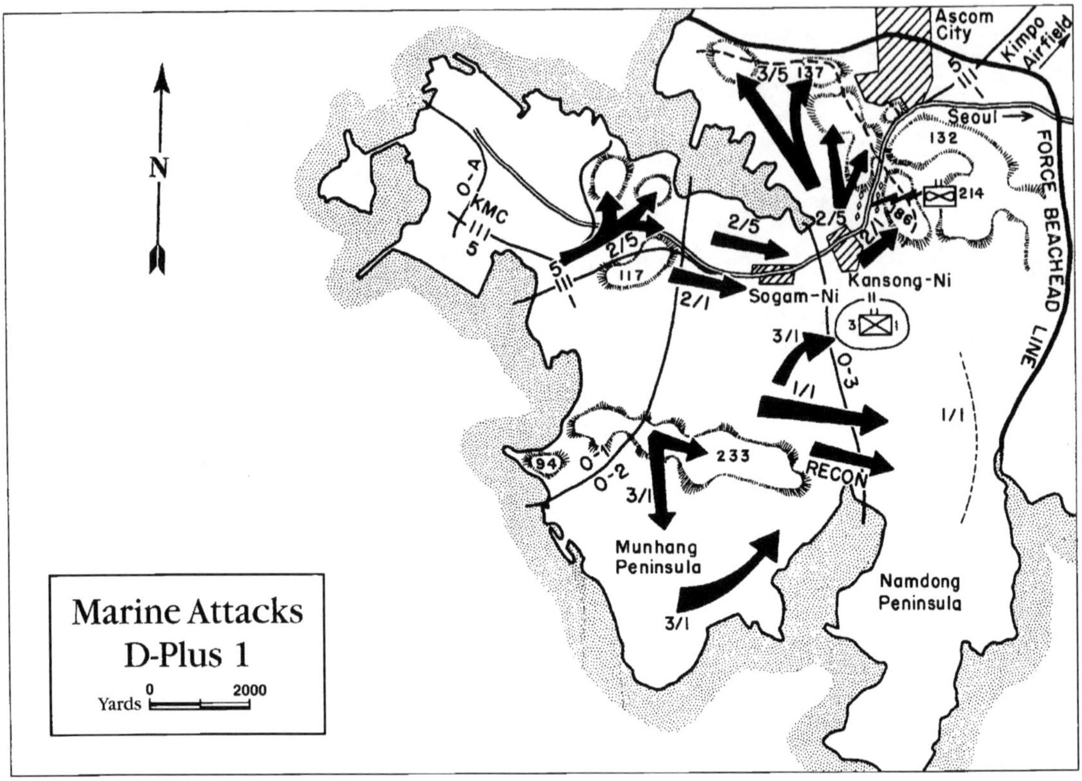

Marine Attacks D-Plus 1

Beach Two, moved in his free boat to explore the possibilities of Blue Beach Three. He found a mud ramp broken through the seawall and some of his battalion was diverted to this landing point. An enemy machine gun in a tower about 500 yards inland caused a few casualties before it was knocked out by fire from the Army's armored tractors.

More serious problems confronted Lieutenant Colonel Jack Hawkins' 1st Battalion, which was in regimental reserve. Boated in LCVP landing craft, he was ordered by Puller, who was already ashore, to land his battalion. If things had gone well Hawkins should have beached at about H+45 minutes or 1815. Veering off far to the left in the gloom, his leading waves mistook the wall of the tidal basin for the seawall of Blue Beach Two. Most of Company B and some of Company A had landed before Hawkins could correct the error. Most of those who landed were reboated and sent on to Blue Beach Two. Because of a shortage of boats, however, one platoon was left behind. Marching overland to Blue Beach Two this orphan platoon gilded the lily by picking up a bag of prisoners enroute.

The 3d Battalion, 1st Marines' reserve—Company H under Captain Clarence E. Corley, Jr.—landed across Blue Beaches Two and Three. The 1st Platoon, led by First Lieutenant William Swanson, had the mission of securing the right flank of the bridgehead. Swanson slid his platoon behind Company I and moved against a platoon-sized enemy dug in on Hill 94, which topped the fishhook cape bounding the beachhead on the south. The North Koreans were driven out, but at a cost. Swanson himself was severely wounded in the thigh and evacuated. (Swanson returned to the 3d Battalion in late winter 1951, was wounded in the hand at the end of March, and killed by one of our own mines on 15 May 1951.)

Corley's Company H, less its 1st Platoon, moved into the gap between Companies G and I. The 2d Platoon, Company H, was sent forward at midnight to outpost Hill 233, a mile to the front, got halfway there, to Hill 180, and received permission to stay put for the night.

Generals Almond and Shepherd came in with the ninth wave, along with Admiral Struble, for a looksee at how events were progressing on Blue Beach. Almond's aide, Lieutenant Haig, had come in to

National Archives Photo (USMC) 127-GK-234I-A409339

On the morning following the landing, the Marines marched through Inchon itself against no resistance. Initially, the Marines enjoyed a 10 to one numerical advantage over the mediocre defense force. North Korean resistance stiffened in both numbers and quality as the attack moved inland toward Seoul.

Red Beach on board one of the LSTs. He had with him Almond's personal baggage and the wherewithal to establish a mobile command post including a van fitted out as sleeping quarters and an office. In transit Haig had lost two of the general's five jeeps, swept over the side of the LST in the typhoon. When Haig met up with his boss, Almond's first question was whether Haig had gotten his baggage ashore without getting it wet.

While the 5th Marines were assaulting Red Beach, Brigadier General Craig—with his brigade dissolved and now the assistant division commander—came ashore at Wolmi-do and, joining Taplett's 3d Battalion, established an advance division command post. Craig had brought his brigade staff ashore intact to function as an interim division staff. Since his arrival in the objective area, Craig had had no opportunity to meet with O. P. Smith face-to-face.

During the night, Taplett's battalion crossed the causeway from Wolmi-do and rejoined the main body of the 5th Marines on Red Beach. Before morning the 1st Marine Division had all its first day's objectives. Resistance had been scattered—of the sort that goes down in the situation report as "light to moderate." Total Marine casualties for the first day's fighting were 20 killed, 1 died of wounds, 1 missing in action, and 174 wounded.

Assault Continues

At about midnight Puller and Murray received the division's attack order for the next day. Murray was to bring the 5th Marines up on line abreast of Puller's 1st Marines. The axis for the advance on Seoul would be the intertwined highway and railroad. The Korean Marine regiment

Department of Defense Photo (USA) SC348506

Once ashore at Inchon, the Marines see for themselves that naval gunfire had destroyed much of the city. Once ashore, the rule-of-thumb was that each assault battalion would have a cruiser or destroyer available for on-call missions.

was initially left behind in Inchon to mop up.

The day, 16 September, was clear and pleasant. The climate was about the same as our northeastern states at this time of year, warm during the day, a bit cool at night.

Murray elected to advance in column of battalions, leading off with Roise's 2d Battalion, followed by the 1st and 3d Battalions in that order. The 2d Battalion's advance through Inchon was strangely quiet. The enemy had vanished during the night.

Corsairs Against T-34s

Five miles to Murray's front, six of the vaunted Soviet-built T-34 tanks, without infantry escort, were rumbling down the Seoul highway toward him. Near the village of Kansong-ni, eight Corsairs from VMF-214 swept down on the advancing tanks with rockets and napalm. One Corsair, flown by Captain William F. Simpson, Jr., failed to come out of its dive, killing Simpson, but the tank attack was halted. One T-34 was engulfed in flames, a second had its tracks knocked off, and a third stood motionless on the road. A second flight of Corsairs came over to finish off the disabled T-34s. The pilots pulled away, thinking incorrectly that all six tanks were dead.

On the ground, Roise's 2d Battalion, 5th Marines, made solid contact with Sutter's 2d Battalion, 1st Marines, on Hill 117. The two battalions continued the advance against nothing heavier than sniper fire. By 1100 elements of both battalions were just short of Kansong-ni where they could see the smoke still rising from the fires set by the battle of T-34s and Corsairs.

Meanwhile, General Craig had moved his command group into Inchon itself. On the outskirts of the city, he found what he thought would be a good location for the division command post including a site close by where a landing strip could be bulldozed. He ordered his temporary command post moved forward.

Thirty "SCAJAP" LSTs, manned for the most part with Japanese crews, had been collected for the Inchon landing. Those that were carrying troops did not beach, but sent their passengers off in amphibian tractors. After the assault waves had swept ahead they did beach, when the tides permitted, for general unloading. Beach conditions and the mixed

The rubber-tired amphibious DUKW pulls a trailer about a mile outside of Inchon on the first morning after the landing. These "ducks" were used primarily to move guns, ammunition, and supplies for the artillery.

Department of Defense Photo (USA) SC348502

Department of Defense Photo (USA) SC348504

A curious Marine passes three knocked-out T-34 tanks. The vaunted Soviet-built tank proved no match for the array of weapons that the Marines could bring to bear, ranging from a Corsair fighter-bomber's rockets to the 3.5-inch rocket launchers in the rifle and weapons companies.

quality of the Japanese crews threw the planned schedule for unloading completely out of balance.

The landing and employment of tanks presented problems. The Marines had just received M-26 Pershings as replacements for their M-4A3 Shermans. Few of the members of Lieutenant Colonel Harry T. Milne's 1st Tank Battalion—except for Company A, which had been with the 5th Marines and had the M-26 at Pusan—were familiar with the Pershing. The tankers received their instruction on the new tanks on board ship—not the best place for tank training.

Major Vincent J. Gottschalk's VMO-6, the division's observation squadron, began flying reconnaissance missions at first light on D+1, 16 September. VMO-6 possessed eight Sikorsky HO3S-1 helicopters

M-26 Pershing tanks emerge from the maws of beached LSTs ("landing ships, tank") at Inchon. Marine tankers, previously equipped with the obsolescent M-4 Sherman tank, were re-equipped with the Pershings literally while on their way to the objective area.

Department of Defense Photo (USMC)

Department of Defense Photo (USMC) A130235

Maj Vincent J. Gottschalk, commanding officer of VMO-6, prepares to take off in an OY light observation aircraft. Among the varied missions of the squadron was spotting and adjusting artillery fire on the retreating North Koreans for the ground troops.

and eight OY airplanes and had been with the 1st Brigade at Pusan where for the first time Marines used helicopters in combat. That day, First Lieutenant Max Nebergall pulled a ditched Navy pilot out of Inchon harbor in the first of many rescue operations.

In the 1st Marines' zone of action Puller sent Ridge's 3d Battalion to make a sweep of Munhang Peninsula. Ridge used amphibian tractors as personnel carriers—a bold but dangerous practice—and advanced on a broad front, Companies G and I abreast with Company H following in reserve. Prisoners and materiel were taken, but there was almost no fighting. By noon the division held the 0-3 line, a front three miles long, secured on both flanks by water. Smith ordered Murray and Puller to move on forward and seize the Force Beachhead Line (FBHL) which would conclude the assault phase of the amphibious operation.

Murray chose to advance in two prongs. Roise with the 2d Battalion would continue to advance with his right flank tied to the Seoul highway. Taplett, coming up from behind with the 3d Battalion was to swing wide to the left. Newton, with the 1st Battalion, would follow in reserve.

Roise's battalion, escorted by Lieutenant Sweet's five M-26 Pershing tanks, moved up the road and at about 1330 rounded the bend into Kansong-ni. Two of Sweet's tanks crawled up a knoll from which they could cover the advancing riflemen. From this vantage point the Marine tankers saw three T-34 tanks, not dead as supposed, but ready for battle with hatches buttoned up and 85mm guns leveled on the bend in the road. Sweet's tanks smacked the T-34s with 20 rounds of armor-piercing shells. The T-34s went up in flames. Company D led the advance past the three burning hulks. Nearby the Marines found the two tanks knocked out earlier by the Corsairs. The sixth tank had vanished.

Company D continued for another thousand yards and then climbed a high hill on the west side of the road. Company F joined Company D on their left. They were still two miles from the Force Beachhead Line, but it looked like a good time and place to dig in for the night.

On Roise's left, Taplett's 3d

Department of Defense Photo (USA) SC349015

For much of the advance up the axis of the Inchon-Seoul highway, and even sometimes traveling cross-country, Marines used amphibian tractors as personnel carriers protected along the way by M-26 tanks. The North Koreans, in turn, tried to choke off these advances with ambushes and antitank mines.

Battalion advanced uneventfully and now held high ground overlooking the FBHL. His patrols reached the edge of Ascom City—once the village of Taejong-ni and now the remnants of a huge service command that had been used by the U.S. Army during the occupation—to his front. The sea was to his left.

A Korean civilian eager to assist the advancing forces, shows one of the division's reconnaissance Marines a large cache of dynamite and ammunition hidden in a storage cave. It was one of several caches uncovered by Capt Kenneth Houghton's Marines on the division's right flank.

Department of Defense Photo (USMC) A2821

South of the 5th Marines, Puller's 1st Marines, having spent most of the day pulling together its scattered parts, did not jump off in the new attack until about 1600. Sutter's 2d Battalion went forward on the right of the road past Kansong-ni for a thousand yards and then tied in with Roise's battalion for the night. Hawkins' 1st Battalion filled in between Sutter and Ridge. Ridge's 3d Battalion had done more hiking than fighting and at the end of the day was relieved by the Division Reconnaissance Company, under pugnacious Captain Kenneth J. Houghton, attached to the 1st Marines as the division's right flank element. Ridge's Marines went into regimental reserve. Houghton's reconnaissance Marines engaged no enemy but found huge caches of arms and ammunition.

O. P. Smith Opens His Command Post

General Craig had just gotten back from his search for a site for the division command post, when he learned that O. P. Smith, accompanied by Admiral Struble and General Shepherd, had landed. Smith was satisfied with Craig's recommended site. Craig then took him for a quick tour of the troop dispositions and at 1800 Smith officially assumed command ashore. During the day, General Almond visited Red Beach and the 5th Marines.

Smith was joined later that evening by Major General Frank E. Lowe, an Army Reserve officer and President Truman's personal observer, who had arrived unannounced. Lowe moved into the division command post. He and Smith got along famously. "His frank and disarming manner made him welcome throughout the division," remembered Smith.

Photo by Frank Noel, Associated Press

Front-line Marines found that stripping a prisoner bare took all the fight out of him and also eliminated the possibility of hidden weapons. Rear-echelon authorities found the practice distasteful and ordered the Marines to desist.

More Enemy T-34 Tanks

The night of 16-17 September was quiet, so quiet, the official history remarks, that a truck coming down the highway from Seoul drove unimpeded through the Marine front lines, until finally stopped by a line of M-26 tanks several hundred yards to the rear. The tankers, the 1st Platoon, Company A, under First Lieutenant William D. Pomeroy, took a surprised NKPA officer and four enlisted men prisoner.

Lieutenant "Hog Jaw" Smith, commander of Company D, 5th Marines, from his observation post overlooking the highway was sufficiently apprehensive, however, about a sharp bend in the road to the left front of his position to outpost it. He dispatched his 2d Platoon with machine guns and rocket launchers attached, all under Second Lieutenant Lee R. Howard, for that purpose.

During the night the North Koreans formed up a tank-infantry column—six T-34s from the *42d NKPA Mechanized Regiment* and about 200 infantry from the *18th NKPA Division* in Seoul—some miles east of Ascom City. Howard saw the lead tank at about dawn, reported its approach to "Hog Jaw" Smith, who reported it to Roise, who could not quite believe it. Obviously the North Koreans did not know the Marines were waiting for them. Howard let the column come abreast of his knoll-top position and then opened up. Official historians Montross and Canzona say: "The Red infantry went down under the hail of lead like wheat under the sickle." Corporal Oley J. Douglas, still armed with the 2.36-inch rocket launcher and not the new 3.5-inch, slid down the hill to get a better shot at the tanks. At a range of 75 yards he killed the first T-34 and damaged the second. The remaining four tanks continued to plow forward to be met by a cacophony of 90mm fire from Pomeroy's M-26 tanks at 600 yards range, 75mm recoilless rifle fire at 500 yards, and more rockets, some coming from Sutter's battalion on the other side of the road. Private First Class Walter C. Monegan, Jr., from Company F, 1st Marines, fired his 3.5-inch rocket launcher at point-blank range. Just which weapons killed which tanks would be argued, but the essentials were that all six T-34s were knocked out and their crews killed.

MacArthur Comes Ashore

MacArthur, instantly recognizable in his braided cap, sunglasses, well-worn khakis, and leather flight jacket, came grandly ashore that same morning, 17 September. His large accompanying party included Struble, Almond,

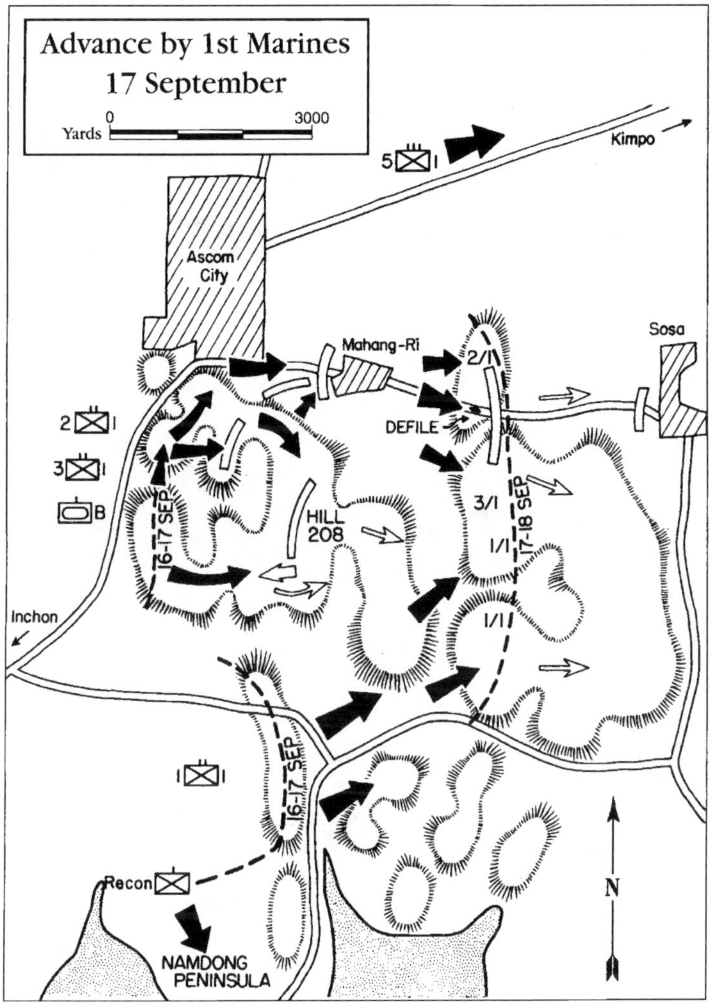

Shepherd, Whitney, Wright, and Fox; a bodyguard bristling with weapons; and a large number of the press corps. A train of jeeps was hastily assembled and the party proceeded to the 1st Marine Division headquarters in a dirt-floored Quonset hut where Smith joined the party. MacArthur presented him a Silver Star medal.

MacArthur and his entourage then visited Puller at the 1st Marines' observation post. MacArthur climbed the hill. Puller put down his binoculars and the two great actors shook hands. MacArthur gave Puller a Silver Star.

MacArthur's cavalcade next drove to the site of the still-smoking hulls of the dreaded North Korean T-34 tanks that had counterattacked at dawn. Shepherd, looking at the still-burning T-34s, commented to Almond that they proved that "bazookas" could destroy tanks.

"You damned Marines!" snorted Almond. "You always seem to be in the right spot at the right time . . . . MacArthur would arrive just as the Marines knocked out five tanks." Shepherd replied, "Well, Ned, we're just doing our job, that's all."

MacArthur climbed back into his jeep and the star-studded party drove on. Seven dazed North Korean soldiers crawled out from the culvert over which MacArthur's jeep had parked and meekly surrendered.

Next stop for MacArthur was the 5th Marines command post. MacArthur went to award Silver Stars to General Craig and Colonel Murray only to learn that his supply of medals was exhausted. "Make a note," he told his aide. The medals were delivered later.

MacArthur finished his tour with a visit to Green Beach at Wolmi-do, where unloading from the LSTs was progressing, and to see the occupants of the prisoner of war stockade—671 of them under guard of the 1st Marine Division's military police.

Ashore at Wolmi-do, MacArthur found evidence, to his great satis-

At a temporary aid station at Pier No. 2, designated Yellow Beach, a wounded Marine is given whole blood by a Navy corpsman. From this station, the wounded were evacuated to hospital ships off shore.

National Archives Photo (USA) 111-SC349024

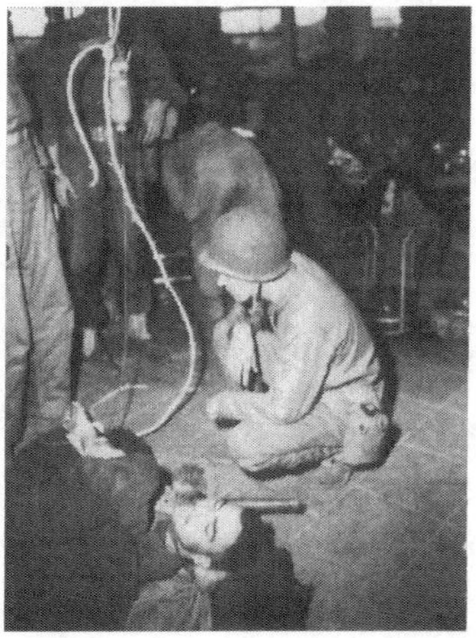

Department of Defense Photo (USA) SC348526

On the morning of 17 September, Gen MacArthur, surrounded by subordinates, bodyguards, and photographers, made a grand and much publicized tour of the Inchon beachhead. MacArthur is unmistakable in his crushed cap, sunglasses, and leather jacket. LtGen Lemuel C. Shepherd, Jr., on the left, is in his usual khakis and carrying his trademark cocomacaque, or Haitian walking stick. MajGen O. P. Smith, in khaki fore-n-aft cap and canvas leggings, trudges along behind Shepherd.

Almost obscured by the jeep's windshield, a photographer peers through his lens at the command echelons of the Inchon landing during the 17 September visit. Gen MacArthur in hawk-like profile stares straight ahead. MajGen O. P. Smith sits smiling in the middle of the rear seat, flanked on his right by MajGen Edward M. Almond and on his left by VAdm Arthur D. Struble. The unidentified Marine driver awaits instructions.

Department of Defense Photo (USA) SC348522

Department of Defense Photo (USA) SC348516

FMFPac commander LtGen Lemuel C. Shepherd, VMI 1917, on the right, points out something significant to the X Corps commander, MajGen Edward M. Almond, VMI 1915, as they move by motor launch from the Mount McKinley *to the beach. Shepherd, although relegated to the position of observer instead of corps commander, held no grudge against Almond.*

faction, that the enemy had begun an intensive fortification of the island. Later he pontificated: "Had I listened to those who wanted to delay the landing until the next high tides, nearly a month later, Wolmi-do would have been an impregnable fortress."

Almond, just before leaving with his boss to return to the *Mount McKinley*, informed Smith that Barr's 7th Infantry Division would begin landing the next day, coming in on the 1st Marine Division's right flank. Smith, returning to his command post, learned that Major General James M. Gavin, USA, of World War II airborne fame, had arrived to study the Marine Corps' use of close air support.

An airstrip was set up next to the division command post that same day, 17 September. After that, Gottschalk's VMO-6 flew a full schedule of observation, evacuation, liaison, and reconnaissance flights.

Marine helicopters, fragile and few in number, were found useful in evacuating severely wounded Marines to hospital facilities to the rear or at sea. As the war progressed, more suitable helicopters arrived and the practice became standard.

Photo by Frank Noel, Associated Press

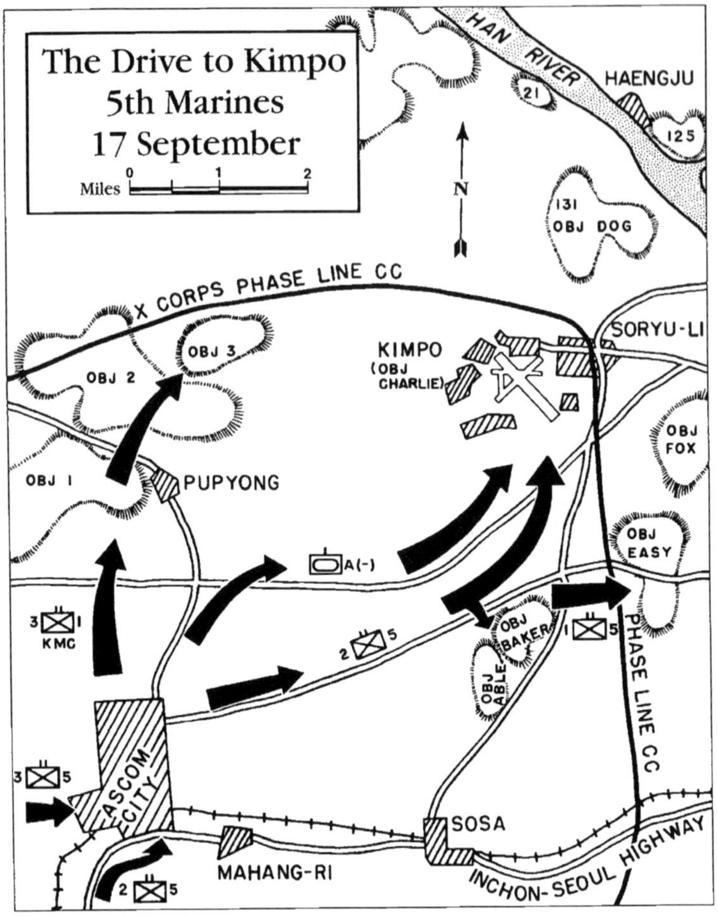

The Drive to Kimpo
5th Marines
17 September

Infantry Advances

The battle with the T-34s delayed for an hour the jump-off for the day's attacks. The next phase line was 19 miles long and Murray's 5th Marines had two-thirds of it. At 0700, the Korean Marines' 3d Battalion had passed through Roise's 2d Battalion to clean up the outskirts of Ascom City. Roise himself jumped off two hours later, Captain Jaskilka's Company E in the lead. The advance was to be in column and then a left turn into Ascom City.

Company E, joined by 2d Platoon, Company F, spent the morning in a methodical clearing of the densely built-up area of little pockets of resistance. Roise found that the road on the map that was supposed to lead to his next objective, four miles distant, was nonexistent on the ground. The renewed advance did not get off until mid-afternoon.

The inexperienced 3d Battalion of the Korean Marines ran into trouble on the other side of Ascom City. Taplett's 3d Battalion, 5th Marines, in regimental reserve, moved in to help and efficiently knocked out the moderate resistance. Pomeroy came up with his platoon of M-26 tanks. Looking for Roise's 2d Battalion and, not finding the mythical road, he instead found Taplett's 3d Battalion. Eventually, Pomeroy reached the 2d Battalion and a road that would lead to Kimpo Airfield now about five miles away. He was joined by his company commander, Captain Gearl M. English, and another platoon of tanks.

Meanwhile, Roise advanced to two high hills some 4,000 yards south of Kimpo. He launched his attack against the airfield with Companies D and E in the assault. They moved rapidly against nothing but light small arms fire. Captain English brought up his tanks to help, assigning a tank platoon to support each of the assault companies. By 1800, Roise's Marines were at the southern end of the main runway. Each of his three rifle companies curled into separate perimeters for the night. Lieutenant Deptula's 1st Platoon, Company E, was positioned well out to the front in the hamlet of Soryu-li as an outpost.

During the afternoon, Newton and the 1st Battalion, 5th Marines, had moved up on Roise's right against no resistance. Taplett's 3d Battalion, having eased the situation for the Korean Marines, was two miles to the rear, again in regimental reserve.

With 1st Marines

Throughout the day, 17 September, Puller's 1st Marines had continued its advance. On the left flank Sutter, with the 2d Battalion, straddled the highway and moved forward behind an intermittent curtain of howitzer fire delivered by the 11th Marines. Essentially, Sutter was attacking due east from Mahang-ri to Sosa, two fair-sized villages. He deployed Company E on the left of the road, Company F on the right, and kept Company D in reserve. As the 5th Marines moved to the northeast toward

Department of Defense Photo (USMC)

When not moving from hill to hill, the Marines frequently found themselves attacking across flat rice paddies. Ironically, Kimpo, in addition to having the best airfield in Korea, was also known for growing the best rice.

Kimpo, a considerable gap widened between the two regiments.

North Korean resistance thickened as Sutter neared Sosa. Puller ordered Ridge to move the 3d Battalion up on Sutter's right flank. Ridge decided again to use amphibian tractors as personnel carriers. Westover's Company G clanked up the road behind the 2d Platoon, Company B tanks, under Second Lieutenant Brian J. Cummings. In a defile, some brave North Koreans tried to stop Cummings' M-26s with grenades. The advance on the road stalled. Company G got up on the high side of the defile to the right of the road. With Sutter's battalion on the left, the Marines had a converging "turkey shoot" and broke up the North Korean attack. Sutter and Ridge dug in for the night, each battalion on its own side of the defile. To their south, Hawkins' 1st Battalion and Houghton's Reconnaissance Company had cleared up Namdong Peninsula. The night would pass quietly for the 1st Marines.

To the rear, Inchon was in a shambles. Most of the city officials had fled before the North Korean capture of the city. Fortunately, Admiral Sohn Won Yil, the chief of naval operations of the ROK Navy, had come ashore to observe the operations of the Korean Marine Regiment. (He also received a MacArthur Silver Star.) Sohn picked a temporary mayor who was installed on the morning of 18 September by authority of a 1st Marine Division proclamation.

5th Marines Takes Kimpo

The night of 17-18 September was tense for the 5th Marines. Murray was certain that the North Koreans would not give up Kimpo, the best airfield in Korea, without a fight, and he was right. The airfield was under the apparent command of a Chinese-trained brigadier general, Wan Yong. The garrison, nominally the NKPA *1st Air Force Division*, was in truth a patchwork of bits and pieces of several regiments, with not more than a few hundred effectives.

The North Koreans went against Roise's well dug-in battalion in

Shore party operations followed close behind the assault waves and within a few days, stocks of ammunition, rations, and other supplies had reached the level needed for the drive to Seoul and its capture.

Photo by Frank Noel, Associated Press

Department of Defense Photo (USN) 420271

Fumigation and bath platoons would arrive later, but during the assault phase of the Inchon operation Marines seized the opportunity to clean up when and where they could. Helmets made convenient washbasins.

three badly coordinated attacks. The first hit Deptula's outpost at about 0300 in the morning, the Communists using rifles and machine pistols, backed by a T-34 tank. Deptula skillfully fought off four half-hearted assaults and by 0500 had withdrawn successfully to Company E's main line of resistance.

The second attack came from both the west and east against Jaskilka's Company E. The third attack hit Harrell's Company F further to the south. Both attacks were easily contained. The routed enemy fled toward the Han River.

At daylight Roise jumped off in pursuit. His Marines swept across the airfield, securing it and its surrounding villages by 1000. Companies E and F mopped up and Company D went on to take Hill 131 overlooking the Han. In 24 hours of fighting, Roise had lost four Marines killed and 19 wounded. His Marines had taken 10 prisoners and had counted about 100 enemy dead.

Correspondents and photographers examine a Russian-built Yak fighter in a destroyed hanger at Kimpo Airfield. Captured by the 2d Battalion, 5th Marines, Marine engineers quickly made the airfield operative and ready to receive elements of MAG-33.

National Archives Photo (USA) 111-SC349036

1st Marines Advances

In the 1st Marines' zone of action, Ridge, with the 3d Battalion outside of Sosa, decided that the center of North Korean resistance must be on Hill 123. During the night he called for naval gunfire. HMS *Kenya*, Captain P. W. Brock commanding, delivered some 300 rounds of 6-inch shells somewhere between Sosa and Hill 123. Ridge's naval gunfire spotter was not sure where they impacted, but Ridge, in the interest of inter-allied cordiality, sent Captain Brock a "well done."

At dawn Sutter charged ahead astride the Seoul highway, Company E on the left of the road and Company D on the right. Premature airbursts on the part of his artillery preparatory fires cost him

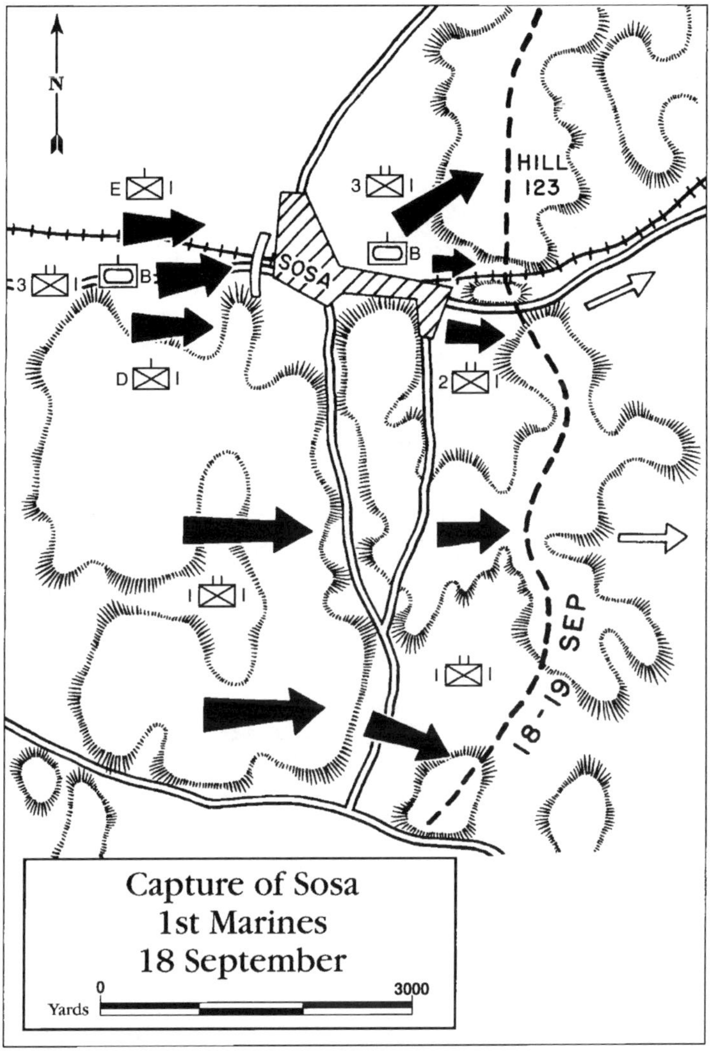

Capture of Sosa
1st Marines
18 September

two killed and three wounded.

Behind the 2d Battalion, Ridge mounted up the 3d Battalion in a motorized column made up of a mixture of jeeps, amphibian tractors (LVTs), and amphibious trucks (DUKWs). Corsairs from VMF-214 worked over Sosa, sighted six T-34s beyond the town, and knocked out two of them. Ridge thundered ahead in a cloud of dust behind the tanks of Company B, 1st Tank Battalion. Together they brushed aside some light resistance, including an antitank roadblock. By noon Ridge had cleared the town. His battalion then swung to the left off the road and moved up Hill 123 while his naval gunfire spotter continued to look for some evidence as to where the *Kenya's* shells might have hit. The 3d Battalion was barely on the hill and not yet dug in when a barrage of North Korean 120mm mortar shells drenched their position causing 30 casualties. The romp over the green hills, marred as they were with the red-orange scars of shell holes and trench lines, was over. The war was getting serious.

Sutter's 2d Battalion, meanwhile, went straight ahead, left flank on the railroad tracks, into a defensive position about a mile beyond Sosa. A barrage of mortar shells cost him 14 casualties. Hawkins' 1st Battalion continued advancing on the right and for the third straight day encountered nothing but a few rifle shots.

Kimpo Airfield Becomes Operational

Murray displaced his command post forward from Ascom City to Kimpo. His regiment spent a quiet day sending patrols around the airfield. The field was in relatively good shape. A North Korean Soviet-built Yakovlev Yak-3 fighter and two Ilyushin "Shturmovik" attack aircraft were found in near-flyable condition.

The first aircraft to land at Kimpo was a Marine H03S-1 helicopter. It arrived at 1000 that morning, 18 September, piloted by Captain Victor A. Armstrong of VMO-6 and with General Shepherd and Colonel Krulak as passengers. General Craig who had just arrived by jeep met them.

Captain George W. King's Company A, 1st Engineer Battalion, made the field operational with temporary repairs. Generals Harris and Cushman came in by helicopter that afternoon. On their advice, General Almond authorized the establishment of Marine Aircraft Group 33 (MAG-33) on the field.

Corsairs began to arrive the next day. Harris set up the headquarters of his Tactical Air Command. Two Corsair squadrons, VMF-312 and VMF-212, came in. Night fighter squadron VMF(N)-542, under Lieutenant Colonel Max J.

National Archives Photo (USMC) 127-N-A3727

1stLt John V. Hanes flew in first Marine Corsair to land at Kimpo Airfield. Having taken hits while on a bombing mission, Hanes was grateful that there was a friendly airfield on which to land. BGen Thomas J. Cushman, Assistant Wing Commander of the 1st Marine Aircraft Wing, greets him.

Volcansek, Jr., arrived from Japan. There was a paper shuffle of squadrons between Marine Aircraft Groups 12 and 33. Marine Aircraft Group 33 under Brigadier General Thomas J. Cushman was now in business ashore. MAG-12 picked up the squadrons afloat. VMFs 214 and 323 continued to operate from the *Sicily* and *Badoeng Strait*, and the night-fighters of VMF(N)-513 from their base at Itazuke in Japan.

Reinforcements Arrive

On Murray's left the 2d KMC Battalion joined the 1st KMC Battalion. The ROK Army's 17th Regiment landed at Inchon and, temporarily under 1st Marine Division control was given an initial mission of completing the clean-up of the unswept area between Ascom City and the sea.

Almond, pressing forward, conferred with Smith on the morning of 18 September concerning the readiness of the 1st Marine Division to cross the Han. Smith pointed out that the 7th Division must take over its zone of action and free his right flank so he could concentrate his forces to cross the river. Smith already had it in his mind that the 5th Marines would go over first to be followed by the 1st Marines. His 7th Marines was still at sea. He went forward to Kimpo to discuss the matter with Murray.

The first unit of the 7th Division, the 32d Infantry, landed, as promised on the 18th, was attached temporarily to the 1st Marine Division. Smith relayed Almond's orders to the 32d to relieve the 1st Marines on the right flank and then to operate in the zone of action assigned to the 7th Division.

7th Division Becomes Operational

On the morning of 19 September, General Barr established his 7th Division's command post ashore. Almond called Barr and Smith together at the 1st Marine Division command post to discuss the 7th Division's immediate assumption of what had been the 1st Marines' zone of action south of the Inchon-Seoul highway.

The 31st Infantry had begun landing. The 32d Infantry would be detached from the 1st Marine Division at 1800. With these two regiments Barr was to begin operations. Smith would then be able to side-slip Puller's regiment fully to the left of the Seoul highway.

Almond's aide, Lieutenant Haig, who was a fly on the tent wall at these meetings, observed that "the Marines' respect for the 7th Division at this stage of the war was ostentatiously low."

Advancing to the Han

After that meeting, the peripatetic Almond went on to visit the command posts of both the 32d Infantry and the 1st Marines. He then proceeded to the 5th Marines command post on Kimpo Airfield to discuss with Murray the crossing of the Han that was scheduled for the following day. Murray told him that he planned to cross in column of battalions using amphibian tractors, amphibious trucks, and pontoon floats at a ferry crossing site northeast of Kimpo.

A significant range of hills separated the 5th Marines on Kimpo from Yongdung-po and the Han. During the night of 18-19 September, Murray had ordered Newton forward with the 1st Battalion to seize Hill 118 and then Hills 80 and 85, overlooking the Kalchon River near where it joined the Han.

At dawn, before Newton could move out, a company-sized North Korean force attacked Company C behind a shower of mortar shells. While Company C slaughtered the North Koreans, "Ike" Fenton's Company B moved against Hill 118. There was the usual air and

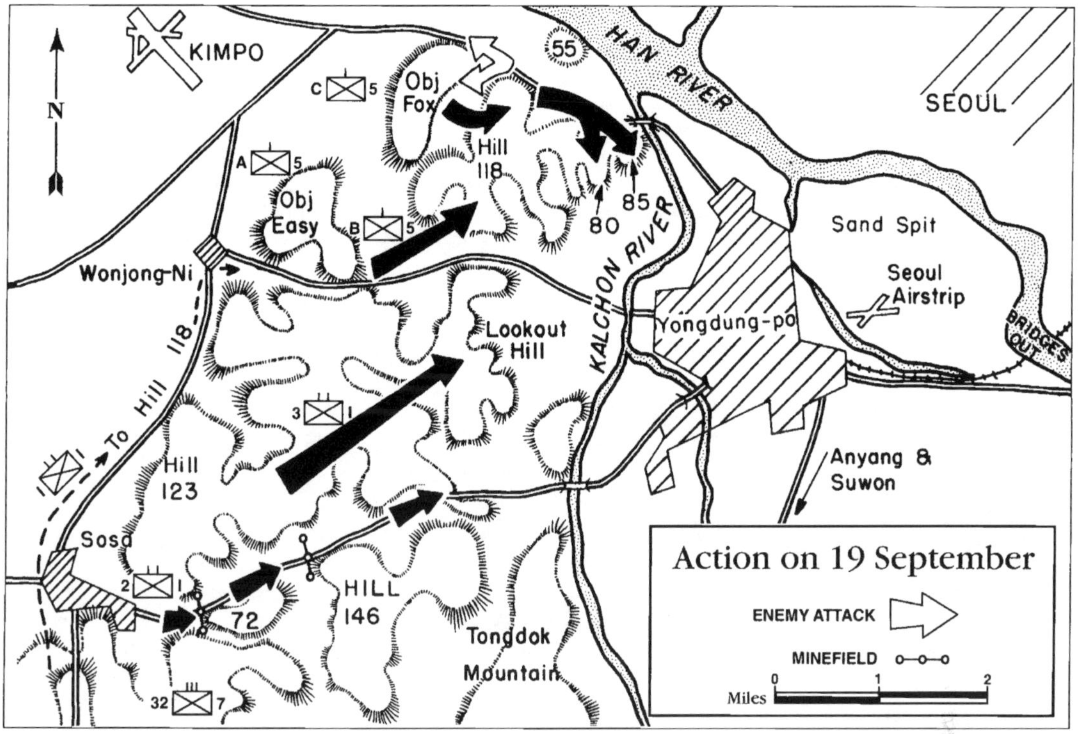

artillery preparation before the jump-off, and Company B took the peak of Hill 118 without suffering a single casualty. The trapped attacking North Koreans lost perhaps 300 dead (there is always optimism in the count of enemy dead) and 100 prisoners. Company C lost two killed and six wounded.

To the 5th Marines' right, Ridge's 3d Battalion, 1st Marines, with Companies H and I in the assault, moved off Hill 123 toward Lookout Hill, so-called because it gave a good view of the Kalchon and the town of Yongdung-po beyond. Official historians Montross and Canzona called the attack, which cost two killed and 15 wounded, "too successful," because it put Ridge's battalion well out in front of the 5th Marines on his left and Sutter's 2d Battalion on his right.

Sutter's battalion was advancing along the Seoul highway behind Captain Richard M. Taylor's Company C tanks and had gone little more than a quarter-mile when the lead M-26 hit a box mine that blew off a track and two road wheels. The antitank barrier of mines was formidable. The whole column came to a stop. Small-arms fire smashed in from neighboring Hill 72. The 11th Marines, the division's artillery regiment, took Hill 72 under howitzer fire. Corsairs from ever-ready VMF-214 came to help. A platoon of engineers under First Lieutenant George A. Babe blew up the box mines with "snowball" charges of C-3 plastic explosive. Sutter used all three of his rifle companies to uncover the minefield and force his way through. His infantry went forward a mile into heavy fighting around Hill 146 while the tanks waited on the side of the road. A second minefield was encountered, and more work by the engineers was needed. At 1900, Sutter ordered his battalion to dig in. His Marines had advanced nearly three miles at a cost of four killed and 18 wounded. Yongdung-po was still more than two miles in front of him.

Smith moved his command post forward the afternoon of 19 September to a site Craig selected about a mile and a half southeast of Kimpo; it had been used for U.S. dependents housing during the occupation. From here Smith was within easy jeep or helicopter distance of his front-line units. The abandoned Quonset huts were near ideal except for occasional harassment apparently by a single NKPA gun. The backbone for the perimeter defense around the command post was provided by a section of the Division Band trained as a machine gun platoon.

The 32d Infantry, now detached

from the division, was somewhere to Sutter's right rear. The Army battalion that relieved Hawkins' battalion had spent the day mopping up rather than continuing the attack.

Hawkins' 1st Battalion, 1st Marines, was on its way to relieve Newton's 1st Battalion, 5th Marines, an 11-mile motor march from the division's right flank. Captain Robert H. Barrow's Company A, 1st Marines, was the first to reach Hill 118 and relieve Fenton's Company B, 5th Marines.

Company C, 1st Marines, was to replace Company C, 5th Marines, on Hills 80 and 85. Newton was anxious to pull back his 1st Battalion, 5th Marines, to Kimpo to get ready for the river crossing the next day, and it was almost dark when Hawkins reached him. Company C, 1st Marines, under Captain Robert P. Wray, had not yet arrived.

Barrow, a tall Louisianian and a future Marine Commandant, realized the tactical importance of Hills 80 and 85 and radioed for permission to move Company A forward to the two hills. Permission was denied. Newton made it known that he would pull Company C off the hills no later than 2100. Wray's Company C did not reach Hill 118 until 2200; Hills 80 and 85 were left empty.

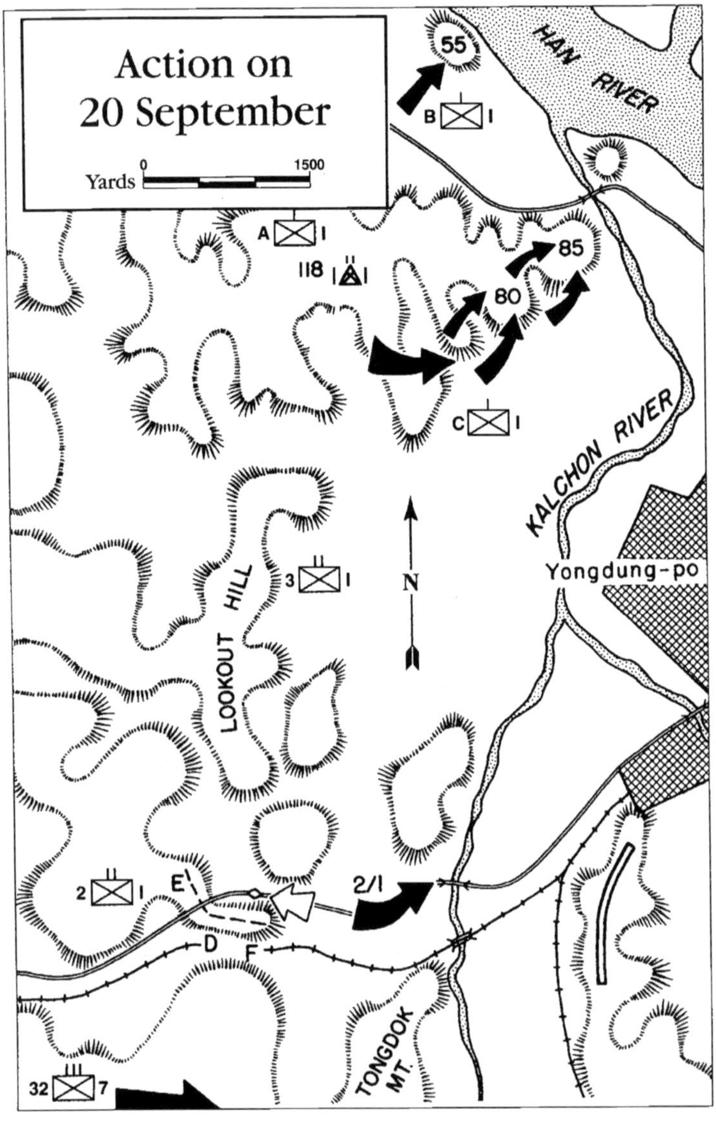

Confused Day

Before dawn the next day, 20 September, Hawkins' Marines on Hill 118 heard the North Koreans assault the empty hills. Then they came on in company-sized strength in a futile attack against the entrenched Marines on Hill 118.

Meanwhile, shortly before dawn a battalion-sized North Korean force, led by five T-34 tanks followed by an ammunition truck, came down the Seoul highway against Sutter's 2d Battalion, 1st Marines. Companies D and E held positions on each side of the road. The column roared through the gap between them and hit head-on against Company F's support position. The North Koreans were caught in a sleeve. Companies D and E poured fire into their flanks. Howitzer fire by the 2d and 4th Battalions, 11th Marines, sealed in the entrapped North Korean column. "A fortunate grenade was dropped in the enemy ammunition truck and offered some illumination," noted the 2d Battalion's Special Action Report, "enabling two tanks to be destroyed by 3.5" rocket fire."

The rocket gunner was Private First Class Monegan, the tank-killer

Private First Class Walter C. Monegan, Jr.

Department of Defense Photo (USMC) A45432

Nineteen-year-old Walter Monegan in five days of action fought two battles against North Korean T-34 tanks, won them both, and lost his own life.

Born on Christmas Day 1930, he could not wait until his 17th birthday, enlisting in the Army in November 1947. The Army discovered he was underage and promptly sent him home. He tried again on 22 March 1948, enlisting in the Marine Corps. After recruit training at Parris Island in June he was sent to China to join the 3d Marines at Tsingtao. After a year in China he came home, was stationed at Camp Pendleton for a year, and then was sent to Marine Barracks, Naval Air Station, Seattle. He had barely re-enlisted in July 1950 when he was ordered to return to Camp Pendleton to join the 2d Battalion, 1st Marines, then being formed.

His remains, buried temporarily at Inchon, were returned home and re-interred in Arlington National Cemetery on 19 July 1951. His wife, Elizabeth C. Monegan, holding their infant child, Walter III, received his posthumous Medal of Honor from Secretary of the Navy Dan Kimball, on 8 February 1952.

Citation:

> For conspicuous gallantry and intrepidity at the risk of his life above and beyond the call of duty while serving as a Rocket Gunner attached to Company F, Second Battalion, First Marines, First Marine Division (Reinforced), in action against enemy aggressor forces near Sosa-ri, Korea, on 17 and 20 September 1950. Dug in a hill overlooking the main Seoul highway when six enemy tanks threatened to break through the Battalion position during a pre- dawn attack on 17 September, Private First Class Monegan promptly moved forward with his bazooka under heavy hostile automatic weapons fire and engaged the lead tank at a range of less than 50 yards. After scoring a direct hit and killing the sole surviving tankman with his carbine as he came through the escape hatch, he boldly fired two more rounds of ammunition at the oncoming tanks, disorganizing the attack and enabling our tank crews to continue blasting with their 90-mm guns. With his own and an adjacent company's position threatened by annihilation when an overwhelming enemy tank-infantry force by-passed the area and proceeded toward the battalion Command Post during the early morning of September 20, he seized his rocket launcher and, in total darkness, charged down the slope of the hill where the tanks had broken through. Quick to act when an illuminating shell hit the area, he scored a direct hit on one of the tanks as hostile rifle and automatic weapons fire raked the area at close range. Again exposing himself he fired another round to destroy a second tank and, as the rear tank turned to retreat, stood upright to fire and was fatally struck down by hostile machine-gun fire when another illuminating shell silhouetted him against the sky. Private First Class Monegan's daring initiative, gallant fighting spirit and courageous devotion to duty were contributing factors in the success of his company in repelling the enemy and his self-sacrificing efforts throughout sustain and enhance the highest traditions of the United States Naval Service. He gallantly gave his life for his country.

of Soryu-li. He slid down the slope from Company F with his 3.5-inch rocket launcher and knocked out the first and second tanks. Machine gun fire killed him as he took aim on the third T-34. His family would receive a posthumous Medal of Honor. A third T-34 was captured intact. Sutter's battalion claimed 300 enemy dead. Half an hour after breaking up the North Korean attack, the 2d Battalion moved forward in its own attack.

Yongdung-po was drenched that day with shell-fire. Puller moved to align his regiment for the assault of the town. Hawkins was to take Hills 80 and 85. Sutter was to advance to the first of two highway bridges crossing the Kalchon. Ridge was to stay in reserve on Lookout Hill.

Hawkins sent out Captain Wray with Company C to capture Hills 80 and 85, that had been free for the taking, the day before. Wray, covered by the 81mm mortars and Browning water-cooled machine guns of Major William L. Bates, Jr.'s Weapons Company, made a text-

Second Lieutenant Henry A. Commiskey

Department of Defense Photo (USMC) A43766

Lieutenant Commiskey was no stranger to war. As an enlisted Marine he had been wounded at Iwo Jima and received a letter of commendation for "exhibiting high qualities of leadership and courage in the face of a stubborn and fanatical enemy."

Born in Hattiesburg, Mississippi, in 1927, he had joined the Marine Corps two days after his 17th birthday. He served more than five years as an enlisted man and was a staff sergeant drill instructor at Parris Island when he was selected for officer training in 1949. He completed this training in June 1950. Two months later he was with the 1st Marines and on his way to Korea.

He came from a family of fighters. His father had been a machine gun instructor in World War I. One brother had been with the Marine Raiders in World War II. Another brother was badly wounded while with the 187th Airborne Infantry in Korea.

In the action on 20 September, that gained Henry Commiskey the nation's highest award for valor, he escaped unscathed, but a week later he was slightly wounded in the fight for Seoul and on 8 December seriously wounded in the knee at the Chosin Reservoir. Sent home for hospitalization, he recovered and went to Pensacola in September 1951 for flight training, receiving his wings in June 1953 and then qualifying as a jet pilot.

He returned to Korea in April 1954 as a pilot with VMA-212. Coming home in September, he returned to line duty at his own request and was assigned once more to the 1st Marine Division. Next assignment was in 1956 to Jackson, Mississippi, close to his birthplace, for three years duty as a recruiter. In 1959, now a major, he went to the Amphibious Warfare School, Junior Course, at Quantico, and stayed on as an instructor at the Basic School. He retired from active duty in 1966 to Meridian, Mississippi, and died of a self-inflicted gunshot wound on 15 August 1971.

Citation:

> For conspicuous gallantry and intrepidity at the risk of his life above and beyond the call of duty while serving as a Platoon Leader in Company C, First Battalion, First Marines, First Marine Division (Reinforced), in action against enemy aggressor forces near Yongdungp'o, Korea, on 20 September 1950. Directed to attack hostile forces well dug in on Hill 85, First Lieutenant Commiskey, then Second Lieutenant, spearheaded the assault, charging up the steep slopes on the run. Coolly disregarding the heavy enemy machine-gun and small-arms fire, he plunged on well forward of the rest of his platoon and was the first man to reach the crest of the objective. Armed only with a pistol, he jumped into a hostile machine-gun emplacement occupied by five enemy troops and quickly disposed of four of the soldiers with his automatic pistol. Grappling with the fifth, First Lieutenant Commiskey knocked him to the ground and held him until he could obtain a weapon from another member of his platoon and kill the last of the enemy gun crew. Countinuing his bold assault, he moved to the next emplacement, killed two or more of the enemy and then led his platoon toward the rear nose of the hill to rout the remainder of the hostile troops and destroy them as they fled from their positions. His valiant leadership and courageous fighting spirit served to inspire the men of his company to heroic endeavor in seizing the objective and reflect the highest credit upon First Lieutenant Commiskey and the United States Naval Service.

Department of Defense Photo (USN) 80-G-426159

On 20 September, as the loading continues, an LST, beached until the next high tide comes in, has discharged its cargo. The small landing craft to the right are a 36-foot LCVP and two 50-foot LCMs.

book double envelopment of Hill 80 against stubborn resistance. The 1st Platoon, under Second Lieutenant William A. Craven, came in on the left. Second Lieutenant Henry A. Commiskey came in on the right with the 3d Platoon. Together they took Hill 80. The day was almost done but Wray went on against Hill 85, repeating his double envelopment. Craven set up a base of fire with his platoon on the northern slope of Hill 80. Second Lieutenant John N. Guild went forward on the left

Amphibious trucks, "ducks" to Marines, are readied at Inchon to be moved up for use in crossing the Han River. The division was well supported by the versatile trucks of the 1st Amphibian Truck Company, an element of the 1st Motor Transport Battalion.

National Archives Photo (USA) 111-SC348700

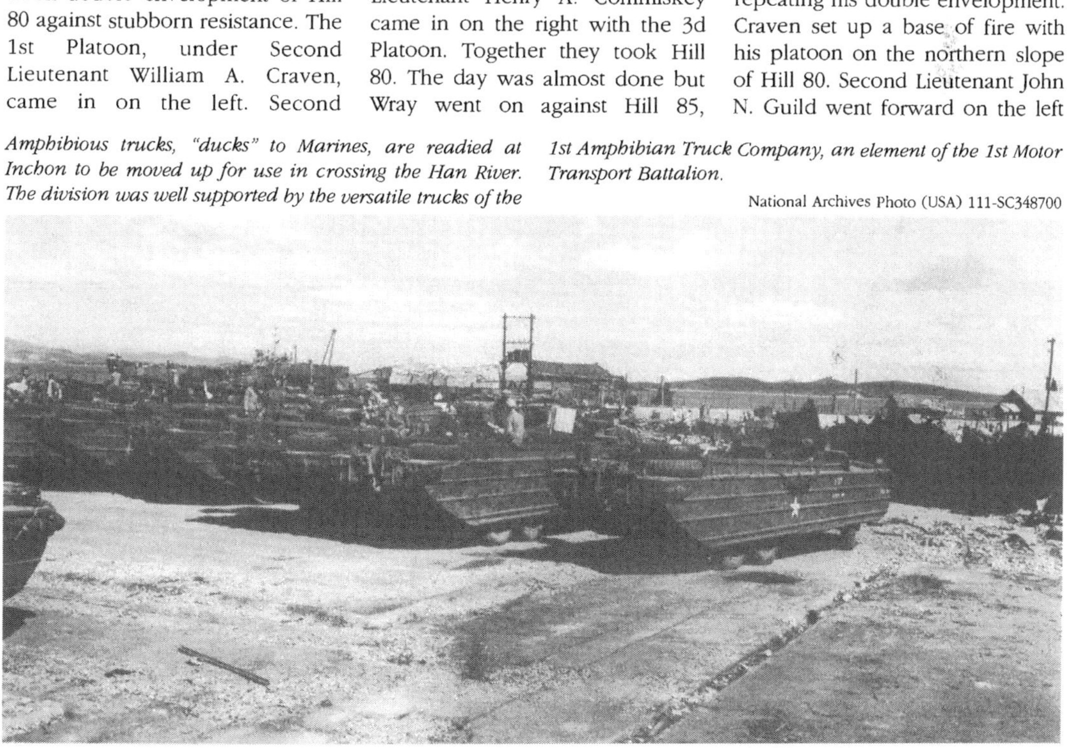

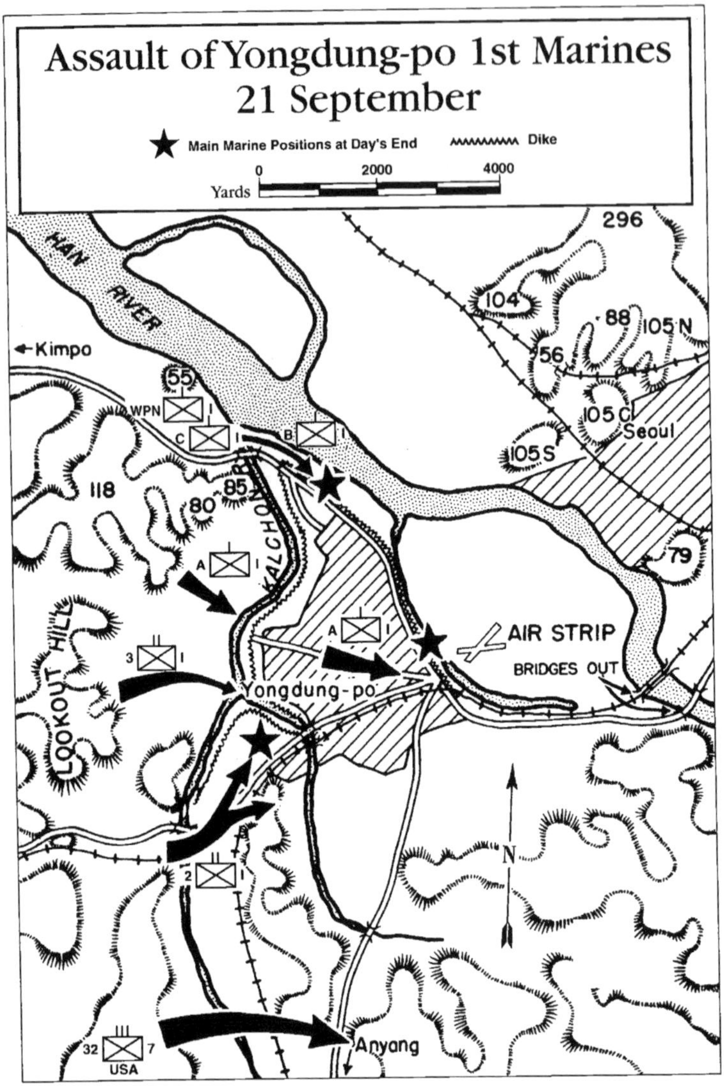

with his 2d Platoon and got almost to the top of the hill before being mortally wounded. Commiskey went out in front of his 3d Platoon in a one-man assault that earned him a Medal of Honor.

While Wray worked at capturing Hills 80 and 85, Hawkins' command group and Barrow's Marines watched as spectators from Hill 118. They saw to their left front, to their horror, a tracked "Weasel" with a wire party from the 1st Signal Battalion hit a mine on the approach to a bridge across the Kalchon near where it joined the Han. In full sight of Hill 118, two Marine wiremen were taken prisoner. A truck from Company A, 1st Engineers, with a driver and three passengers, unaware of the fate of the communicators, now came along the road. Barrow tried to catch their attention with rifle fire over their heads, but the truck continued into the ambush. Three engineers got away; one, Private First Class Clayton O. Edwards, was captured. (He would later escape from a train taking prisoners into North Korea.)

Meanwhile, Sutter's 2d Battalion, having begun the day by breaking up the T-34 tank-led North Korean attack, had moved forward uneventfully, except for harassing fire from their open right flank. They reached their day's objective, the highway bridge over the Kalchon, shortly after noon. The bridge was a long concrete span. The engineers inspected it and certified it strong enough to bear M-26 Pershing tanks for next day's attack into Yongdung-po itself. The second bridge, crossing a tributary of the Kalchon, lay 2,000 yards ahead. A high ridge, seemingly teeming with North Korean defenders, to the right of the road dominated the bridge. Sutter's neighbor on his right was Lieutenant Colonel Charles M. Mount, USA, with the 2d Battalion, 32d Infantry. The ridge commanding the second bridge was technically in Mount's zone of action. At 1300, Sutter asked Mount for permission to fire against the ridge. Mount readily agreed, but it took seven hours to get the fire mission cleared through the layers of regimental and division staffs and approved by X Corps. It was dark before Colonel Brower's 11th Marines was allowed to fire.

During the day, General Almond visited Colonel Puller at the 1st Marines' command post. Almond admired Puller's aggressive tactics and there was also a Virginia Military Institute connection. Puller, saying he could not reach Smith either by wire or radio, asked permission to burn Yongdung-po before committing his troops to its capture. Almond authorized its burning.

Almond's habit of visiting the

Marine regiments and issuing orders directly to subordinate commanders had become a serious aggravation to Smith. A division order went out that any direct order received from Almond would be immediately relayed to division headquarters for ratification.

Ready to Cross the Han

The shelling of Yongdung-po, now blazing with fires, continued throughout the night. Puller's plan of attack for the 1st Marines on 21 September was to have the 2d Battalion continue its advance astride the Inchon-Seoul highway. The 1st Battalion on the left would attack across country and the 3d Battalion, occupying Lookout Hill, would initially stay in reserve.

During the previous day, Captain Richard L. Bland had occupied Hill 55 overlooking the Han with Company B, 1st Marines. Now, shortly after dawn, he took his company across the bridge that had been the site of the ambush of the Marine communicators and engineers. In late afternoon, Hawkins sent Company C and Weapons Company across the bridge to join with Company B to form a perimeter for the night.

During the day, Ridge's 3d Battalion, in reserve on Lookout Hill, had grown impatient and had come forward prematurely, getting out in front of both the 1st and 2d Battalions. Its prospective assault companies, Companies G and I, reached and huddled behind the dike on the western bank of the Kalchon close to a water gate where a tributary entered into the main stream. This put them in good position to watch the approach march of Barrow's Company A to the Kalchon.

With Bland's Company B stalled on the opposite bank of the Kalchon, Hawkins had committed Company A to an attack from its positions on Hill 80 across a mile of rice paddies to the river. Barrow

Following a burst of sniper fire, Marines quickly take cover along a dike near the Han River. So far, the Marines had suffered only light casualties, while the North Koreans had lost heavily.

National Archives Photo (USA) 111-SC349026

National Archives Photo (USMC) 127-N-A3610

Marines of Capt Robert H. Barrow's Company A, 1st Battalion, 1st Marines, move into Yongdung-po. Although the town seemed empty and dead, they carefully searched each building and side street but failed to uncover a flicker of enemy resistance.

deployed his platoons in a classic two-up one-back formation. As they came forward through the waist-high rice straw, a 3d Battalion officer, watching from his position behind the dike, was reminded of the stories he had been told of the Marines advancing through the wheat into Belleau Wood. Without a shot being fired, Company A waded the stream and marched into Yongdung-po. Barrow radioed Hawkins for instructions. Hawkins told him to keep on going.

The crossing of the Kalchon by Ridge's 3d Battalion was less easy. Going over the dike was eerily like going "over the top" of the trenches in the First World War. Second Lieutenant Spencer H. Jarnagin of Company G formed his platoon in line on the near side of the dike close to the water gate. At his whistle signal they started across. As they came out of the defilade provided by the dike, Maxim heavy machine guns on the opposite dike, perhaps 50 yards distant, opened up. Jarnagin fell back dead. His platoon recoiled, some of them wounded. Denied artillery support and with his 81mm mortars lacking ammunition, the battalion's Weapons Company commander called up his platoon of six water-cooled Browning machine guns.

During the rapid cross-country movement toward Seoul the heavy machine guns were initially attached by section to the rifle companies. They could not keep up with the light machine guns nor did the rifle company commanders fully understand their capabilities. Consequently they were pulled back to company control and employed in battery for overhead fire in the attack. Now, in this situation so much like the Western Front, they would come into their own.

With their barrels just clearing the top of the dike, the Brownings engaged the Maxims, just as they had done in 1918, and it was the Brownings that won. The 3d Battalion then crossed the Kalchon at the water gate, Westover's Company G to the left of the tributary, First Lieutenant Joseph Fisher's Company I to the right.

Early that morning Sutter's battalion crossed the second bridge without incident except for fire that continued to come in from across the boundary separating the 1st Marine Division from the 7th Infantry Division. Frustrated by the lack of artillery support, Sutter seized the bit in his teeth and shelled the offending ridge with his attached 4.2-inch mortars before sending up Companies E and F to take the high ground. While they were so engaged, Captain Welby W. Cronk took Company D along the highway and ran into another section of heavily fortified dike. Heavy fighting, supported by the ever-willing Corsairs of VMF-214, continued in Sutter's zone until late in the evening, when Sutter recalled Companies E and F to tuck them into a battalion perimeter for the night.

In Yongdung-po, Barrow could hear the furious firefight being waged by Sutter's battalion somewhere to his right. Crossing the town against scattered opposition Barrow reached yet another dike. Beyond it was a sandy flat reaching about a mile to the Han. To his left rear was Bland's Company B. Barrow dug in on the dike in a sausage-shaped perimeter. At nightfall, the Marines of Company A heard the characteristic chugging clatter of advancing tanks. Five T-34s, without infantry escort, came up the Inchon-Seoul highway and pumped steel into the western face of Company A's position. Barrow's 3.5-inch rocket gunners knocked out one and damaged two others.

Almond had been returning each evening to the *Mount McKinley*, but on the morning of 21 September he moved the headquarters of X Corps ashore and opened his command post in Inchon.

MacArthur came ashore again that afternoon enroute to Japan. A pride of generals—Almond, Shepherd, Smith, Barr, Harris, and Lowe—had gathered at Kimpo Airfield to see him off. Mutual congratulations were exchanged, and MacArthur flew to Tokyo. "He was, in my opinion, the greatest military leader of our century," mused General Shepherd, the Virginia gentleman, in 1967.

Later that day, in a ceremony at X Corps headquarters in Inchon and in accordance with established amphibious doctrine, overall command of the operation passed from Admiral Struble to General Almond.

By midnight, five infantry assaults against Barrow's position had followed the attack by the T-34s. All were beaten back, the heaviest fighting being in front of Second Lieutenant John J. Swords' 3d Platoon.

Pause in the Fighting

When the morning of 22 September came, Barrow's Marines

Gen Douglas MacArthur and MajGen Edward M. Almond examine a map at Kimpo Airfield shortly before the general's departure for Japan. MacArthur would tell the Joint Chiefs of Staff that "his forces were pounding at the gates of Seoul."

National Archives Photo (USA) 111-SC349084

National Archives Photo (USA) 111-SC349054

Navy Hospital Corpsmen Richard E. Rosegoom and Frank J. Yasso, assigned to the 1st Marine Division, give first aid to a wounded North Korean while another prisoner is marched to the rear. While always there for Marines, corpsmen also were available to treat prisoners of war and Korean civilians; the latter were second in number only to the Marines themselves.

were able to count 275 enemy dead. The four remaining T-34s, two damaged, two intact, were found abandoned nearby. The 1st and 3d Battalions renewed their attack and converged on Barrow's position against negligible resistance.

Sutter was not the only commander to complain about the fire control problems along the boundary between the two divisions. The 7th Division reported Marine Corps fire falling in its zone. Almond met with Barr and Smith and then told his aide, Lieutenant Haig, to telephone Corps headquarters and straighten out the situation.

Almond, continuing his critique of the Marines' performance, expressed his concern over Smith's "open" left flank. Smith explained to Almond his use of the 1st Korean Marine Corps Regiment, also that he had formed a Kimpo Airfield defense force, using combat support and service units. Almond appeared somewhat mollified.

The Korean Marines, leaving one battalion behind in Inchon, had followed the 5th Marines to Kimpo Airfield, and made its first attack northwest of the airfield on 19 September against light resistance. That same day the battalion from Inchon rejoined its parent regiment. Now, with one battalion to be left behind to cover the northwest flank, the KMC regiment prepared to follow the 5th Marines across the Han.

Smith's third organic infantry regiment, the 7th Marines, including the battalion that had come from the Mediterranean by way of the Suez Canal, had arrived in the harbor. Colonel Homer Litzenberg asked General Smith what element he wanted landed first. "An infantry battalion," said Smith. "And what next?" "Another infantry battalion."

Litzenberg opened his command

post two miles south of Kimpo. His 3d Battalion, under Major Maurice E. Roach, moved into an assembly area nearby. The 1st Battalion, under Lieutenant Colonel Thornton M. Hinkle, reached Hill 131 a mile north of the airfield sometime after midnight. The 1st Battalion, under Lieutenant Colonel Raymond G. Davis, stayed in the harbor to unload the ships that had brought in the regiment.

Smith made a note in his journal that Almond's concerns over open flanks had increased now that X Corps' command post was ashore. With the arrival of the 7th Marines, Smith himself could rest more easily concerning the security of his northwest flank.

Coordination between the 1st Marine Division and the 7th Division continued to be poor. An extensive minefield delayed the 32d Infantry as it attacked along the Seoul-Suwon highway on 20 September, but on that same day the 32d did take T'ongdok mountain and a part of Copper Mine Hill. The rest of Copper Mine Hill was taken the next day and, as night fell, the Army regiment held a line two miles south of Anyang-ni. The big event of 22 September for the 32d Infantry was the capture of Suwon Airfield and opening it to friendly traffic.

Sutter's 2d Battalion reverted to regimental reserve the afternoon of 22 September after seven days in the assault. His grimy Marines gathered together in a bivouac area where they could wash and rest. The 22 September entry in Almond's war diary, dutifully kept by Haig, noted that Sutter's battalion had taken 116 casualties as "the result of aggressive forward movement without the required artillery preparation." That evening, Almond, after a busy day, entertained Admiral Doyle and selected staff officers at dinner at his newly established mess in Inchon.

Almond and Smith Disagree

By 23 September, the 32d Infantry had secured its objectives

Loaded in amphibious tractors and trucks, Korean Marines prepare to follow the 5th Marines across the Han River. A major portion of the 1st Korean Marine Corps Regiment had followed the 5th Marines to Kimpo Airfield, made its first attack northwest of the field, and were now poised to liberate the Korean capital.

National Archives Photo (USA) 111-SC348702

1st Marine Division Casualties

15-23 September 1950

Date	KIA[1]	DOW[2]	MIA[3]	WIA[4]	Total
15 Sept	20	1	1	174	196
16 Sept	2	1	1	22	26
17 Sept	6	0	0	70	76
18 Sept	7	3	0	92	102
19 Sept	10	1	0	61	72
20 Sept	24	1	3	119	147
21 Sept	30	3	0	198	231
22 Sept	27	3	0	135	165
23 Sept	19	7	0	117	143
Total	145	20	5	988	1,158

1 KIA Killed in Action
2 DOW Died in Action
3 MIA Missing in Action
4 WIA Wounded in Action

overlooking the Han, south and southeast of Yongdung-po. The 3d Battalion of the Army's highly regarded 187th Airborne Regiment, with Almond's "GHQ Raider Group" attached, arrived at Kimpo and temporarily came under 1st Marine Division control. Smith gave it the mission of covering his northwest flank, freeing the 7th Marines for a crossing of the Han.

Almond ordered his command post displaced forward from Inchon to Ascom City. During the day he visited Barr's command post and passed out a liberal number of Silver Stars, Bronze Stars, and Purple Hearts. Smith found Almond's practice of presenting on-the-spot awards disruptive and a cause for hurt feelings and misunderstandings. He thought Almond was inspired by Napoleon, but MacArthur was a more immediate practitioner. Smith had, it will be remembered, himself received a Silver Star from MacArthur as had Barr and Admiral Doyle. MacArthur was even more generous to Admiral Struble, giving him the Army's Distinguished Service Cross.

The 5th Marines was now firmly across the Han but was having difficulty in expanding its bridgehead. Mid-morning on the 23d, Almond met with Smith and urged him to put the 1st Marines across the river. He again complained that the Marines were not pressing the attack vigorously enough. Almond suggested that Smith cross the Han southeast of Seoul with the 1st Marines and then attack frontally into the city. Smith countered with a less-rash plan to have the 1st Marines cross at the 5th Marines' bridgehead. Almond reluctantly concurred.

From 15 through 23 September, the 1st Marine Division had suffered 165 men killed in action or died of wounds, 5 Marines still missing in action, and 988 men wounded. In turn the division had taken, by fairly accurate count, 1,873 prisoners, and claimed 6,500 enemy casualties.

During the day, 23 September, Smith visited the observation post of the 3d Battalion, 1st Marines, which had just taken Hill 108 overlooking the rail and highway bridges, their spans broken, into Seoul. A Marine major, who knew of O. P. Smith's study of the Civil War, presumed to remark that the position was similar to that of Burnside at Falmouth on the north bank of the Rappahannock across from Fredericksburg in December 1862. General Smith looked with amusement at the major and patiently explained that he would not make the same mistake as Burnside. There would be no frontal assault across the river into Seoul.

About the Author

Edwin Howard Simmons, a retired Marine brigadier general, was, as a major, the commanding officer of Weapons Company, 3d Battalion, 1st Marines, in the landing across Blue Beach Two at Inchon. His active service spanned 30 years—1942 to 1972—and included combat in World War II and Vietnam as well as Korea. A writer and historian all his adult life, he was the Director of Marine Corps History and Museums from 1972 until 1996 and is now the Director Emeritus.

He was born in Billingsport, New Jersey, the site of a battle along the Delaware River in the American Revolution, and received his commission in the Marine Corps through the Army ROTC at Lehigh University. He also has a master's degree from Ohio State University and is a graduate of the National War College. A one-time managing editor of the *Marine Corps Gazette*, he has been published widely, including more than 300 articles and essays. His most recent books are *The United States Marines: A History* (1998), *The Marines* (1998), and *Dog Company Six* (2000).

He is married, has four grown children, and lives with his wife, Frances, at their residence, "Dunmarchin," two miles up the Potomac from Mount Vernon.

THIS PAMPHLET HISTORY, one in a series devoted to U.S. Marines in the Korean War era, is published for the education and training of Marines by the History and Museums Division, Headquarters, U.S. Marine Corps, Washington, D.C., as part of the U.S. Department of Defense observance of the 50th anniversary of that war. Editorial costs have been defrayed in part by contributions from members of the Marine Corps Heritage Foundation.

KOREAN WAR COMMEMORATIVE SERIES

DIRECTOR OF MARINE CORPS HISTORY AND MUSEUMS
Colonel John W. Ripley, USMC (RET)
GENERAL EDITOR,
KOREAN WAR COMMEMORATIVE SERIES
Charles R. Smith
EDITING AND DESIGN SECTION, HISTORY AND MUSEUMS DIVISION
Robert E. Struder, *Senior Editor*
W. Stephen Hill, *Visual Information Specialist*
Catherine A. Kerns, *Composition Services Technician*

Marine Corps Historical Center
Building 58, Washington Navy Yard
Washington, D.C. 20374-5040
2000

PCN 190 00315 100

Sources

The official history, *The Inchon-Seoul Operation* by Lynn Montross and Capt Nicholas A. Canzona, Volume II in the series, *U.S. Marine Operations in Korea, 1950-1953* (Washington, D.C.: Historical Branch, G-3 Division, HQMC, 1955), provided a centerline for this account.

Other official histories of great use were Roy E. Appleman, *South to the Naktong, North to the Yalu* (Washington, D.C.: Office of the Chief of Military History, Department of the Army, 1961); James A. Field, Jr., *History of United States Naval Operations: Korea* (Washington, D.C.: Government Printing Office, 1962); and James F. Schnabel, *Policy and Direction: The First Year* (Washington: Office of the Chief of Military History, Department of the Army, 1972).

Victory at High Tide (Philadelphia: Lippincott, 1968) by Col Robert D. Heinl, Jr., remains the best single-volume account of Inchon.

Among the other useful secondary sources were Alexander Haig, Jr., *Inner Circles* (New York: Warner Books, 1992); Clay Blair, *The Forgotten War: America in Korea, 1950-1953* (New York: *Times* Books, 1987); Gen Douglas MacArthur, *Reminiscences* (New York: McGraw Hill, 1964); Gen Omar N. Bradley and Clay Blair, *A General's Life: An Autobiography* (New York: Simon and Schuster, 1983); Donald Knox, *The Korean War: Pusan to Chosin* (San Diego: Harcourt Brace Jovanovich, 1985); Marguerite Higgins, *War in Korea: The Report of a Women Combat Correspondent* (Garden City: Doubleday & Company, 1951); Gen J. Lawton Collins, *Lightning Joe: An Autobiography* (Baton Rouge: Louisiana State University Press, 1979); and Cdr Malcolm W. Cagle and Cdr Frank A. Manson, *The Sea War in Korea* (Annapolis: U.S. Naval Institute, 1957).

Valuable insights were provided by an Inchon war game developed at the Marine Corps Historical Center (MCHC) in 1987, which examined the operation from the viewpoint of its principal commanders, using their reports, writings, and memoirs. Among the primary sources used, the most important were the unit files and records held by MCHC of the 1st Marine Division and its subordinate regiments and battalions. Also important were the biographic files held by Reference Section.

Other primary sources of great use were the oral histories, diaries, and memoirs of many of the participants. The most important of these were those of Generals Stratemeyer, Almond, Cates, Shepherd, O. P. Smith, Craig, V. H. Krulak, and Bowser, and Admirals Burke and Doyle. A fully annotated draft of the text is on file at the Marine Corps Historical Center. As is their tradition, the members of the staff at the Center were fully supportive in the production of this anniversary pamphlet. Photographs by Frank Noel are used with the permission of Associated Press/World Wide Photos.

Made in the USA
Monee, IL
02 December 2025

37199006R00039